Springer Theses

Recognizing Outstanding Ph.D. Research

Aims and Scope

The series "Springer Theses" brings together a selection of the very best Ph.D. theses from around the world and across the physical sciences. Nominated and endorsed by two recognized specialists, each published volume has been selected for its scientific excellence and the high impact of its contents for the pertinent field of research. For greater accessibility to non-specialists, the published versions include an extended introduction, as well as a foreword by the student's supervisor explaining the special relevance of the work for the field. As a whole, the series will provide a valuable resource both for newcomers to the research fields described, and for other scientists seeking detailed background information on special questions. Finally, it provides an accredited documentation of the valuable contributions made by today's younger generation of scientists.

Theses are accepted into the series by invited nomination only and must fulfill all of the following criteria

- They must be written in good English.
- The topic should fall within the confines of Chemistry, Physics, Earth Sciences, Engineering and related interdisciplinary fields such as Materials, Nanoscience, Chemical Engineering, Complex Systems and Biophysics.
- The work reported in the thesis must represent a significant scientific advance.
- If the thesis includes previously published material, permission to reproduce this must be gained from the respective copyright holder.
- They must have been examined and passed during the 12 months prior to nomination.
- Each thesis should include a foreword by the supervisor outlining the significance of its content.
- The theses should have a clearly defined structure including an introduction accessible to scientists not expert in that particular field.

More information about this series at http://www.springer.com/series/8790

Guorui Wang

Characterization and Modification of Graphene-Based Interfacial Mechanical Behavior

Doctoral Thesis accepted by
University of Science and Technology of China,
Hefei, China

Springer

Author
Dr. Guorui Wang
Department of Modern Mechanics
University of Science and Technology
of China
Hefei, China

Supervisors
Prof. Xiaoping Wu
Department of Modern Mechanics
University of Science and Technology
of China
Hefei, China

Prof. Zhong Zhang
National Center for Nanoscience
and Technology
Beijing, China

ISSN 2190-5053 ISSN 2190-5061 (electronic)
Springer Theses
ISBN 978-981-15-8028-4 ISBN 978-981-15-8029-1 (eBook)
https://doi.org/10.1007/978-981-15-8029-1

This Springer imprint is published by the registered company Springer Nature Singapore Pte Ltd.
The registered company address is: 152 Beach Road, #21-01/04 Gateway East, Singapore 189721, Singapore

Supervisors' Foreword

Graphene, being a wonder material for more than a decade, has garnered enormous interests from both academia and industry due to its vast potential to enable the design of materials and devices with novel functionalities, as well as to improve the performances of existing products. While tremendous efforts have targeted electronic transport and optical properties of graphene, the graphene mechanics is highly complementary and could, in turn, interact with the physics. Distinctively, the unique combination of the atomic smoothness and huge specific surface area of graphene results in rich mechanical behaviors hidden at the interface. A full understanding of the interfacial deformation and failure mechanisms in terms of different loading conditions is hence indispensable for graphene-based applications. However, the difficulty in observing and measuring such interfacial behaviors at atomic length scales has been a daunting challenge. Also, the graphene-based nano-interfaces challenge the applicability of continuum mechanics theories developed for thin-film systems.

To address these issues, this thesis provides a comprehensive experimental framework for characterizing graphene interfaces with various materials, including polymers, silicon oxide and 2D materials. Raman spectroscopy, combined with atomic force microscopy, offers a powerful tool to track graphene deformations and enables the quantification of interfacial properties at the nanoscale. The pristine van der Waals interfaces are found to be extremely susceptible to shear sliding, even without normal pressure, which serves as a unique stress transfer mechanism for nano-interfaces. In addition, these slip events are typically the principal reason for triggering of elastic instabilities under complex stress states as highly flexible graphene easily deforms out of plane. Consequently, these buckles behave as the interfacial defects that greatly degrade the fatigue life performance of graphene-based structures. The interplay between the elasticity of graphene and the interfacial shear interactions is further highlighted. An example of a representative is the first measurement of an ultralow interlayer shear stress of 40 kPa between graphene layers via a novel pressurized bulging test. The interlayer shear effect not only accounts for the elastic softening of in-plane stiffness but also leads to an anomalous scaling of bending rigidity that violates the classic plate theory. Beyond these, the thesis also bridged between the surface chemistry and the interfacial mechanics of graphene and laid the foundation for developing graphene-based materials and devices with desired performance.

Overall, Dr. Guorui Wang's thesis opens new chapters in the exploration of the interfaces in complex graphene-based systems, having significant implications in diverse applications such as nanocomposite, flexible electronics and nanoelectromechanical devices. Although this thesis concentrates on graphene, both the observed phenomena and proposed methodology can be directly transferable to other 2D materials and van der Waals heterostructures. These original works could shed light on some significant issues in the field of 2D materials beyond graphene and allow readers to grasp new ideas from a perspective of nanomechanics.

Hefei, China
Beijing, China
June 2020

Prof. Xiaoping Wu
Prof. Zhong Zhang

Abstract

Owing to the unique two-dimensional (2D) crystal structure and the atomic thickness, graphene exhibits unprecedented mechanical performance as well as outstanding electrical, optical and thermal properties, which endow graphene with a bright prospect in the fields of functional composite materials, flexible electronics, nanoelectromechanical systems (NEMS) and so on. Classic view of conventional composites and devices implies that the macroscopic performance is strongly dependent on the interfacial quality and stability. Weak adhesion can result in interfacial failures and subsequent loss of functionality, substantially hampering the reliability and efficacy of graphene-based structures. Such an issue could be more critical when the dimensions are downscaled to the nanometer regime, as the weak non-classical forces, such as van der Waals forces, tend to be dominant. On the other hand, ultrahigh surface-to-volume ratio of graphene creates more interface area and induces high sensitivity to the interfacial interactions. Thus, it is believed that graphene could offer an ideal platform to study novel features of friction and adhesion present only at the nanoscale.

In reality, significant gaps have been presented between experimental observations and theoretical predictions upon mechanical performance of graphene-based systems, primarily due to its lubricated nature that was previously overlooked in the mechanical sense. Even though the atomic thinness and high flexibility bestow graphene with a robust adhesion with the substrate, van der Waals (vdW) interfaces feature a considerably weak tangential resistance, highly susceptible to the shear deformation. Consequently, a bulk of questions arise that are of particular importance to both metrology and functionality applications of graphene. How to precisely characterize nanoscale interfacial deformations and measure the interfacial properties between graphene and substrates? How do the interfacial shear effect modify the mechanical responses and interact with intrinsic mechanical properties of graphene? How to effectively tailor the interfacial properties of graphene? This thesis is aiming at answering these questions concerning the interfacial mechanics of graphene. Comprehensively, we study different types of interfaces in this area with special emphasis on the role of the shear sliding.

The interfacial stress transfer mechanism is first explored between graphene and polymer matrix. Given the rehealable nature of vdW interactions, interfacial shear stress is considered as a constant and modeled by a nonlinear shear-lag model. By monitoring the in-situ deformation of graphene, possible failure modes (e.g., slippage, wrinkling, buckle delamination) are identified at nanoscale interfaces and the interfacial stability is examined in terms of a cyclic loading. Leveraging the tunable surface chemistry of graphene, the effect of interfacial bonding types on the graphene-based interfacial mechanical behaviors is elucidated.

We develop a combined Raman-AFM-bulging technique to monitor the interfacial shear deformation between graphene and SiO_2 substrate. Taking into account a sliding boundary, the modified Hencky's model enables the estimation of the interfacial friction. Such an interfacial shear effect on the elastic stiffness and adhesion properties of graphene membranes is further elucidated. We also discuss in detail how the interfacial shear sliding interweaves the surface instabilities and controls the evolution of bubble morphology. The versatility of the bulging technique is finally demonstrated for nanomechanical characterization of 2D materials beyond graphene.

The interlayer mechanical coupling determines the physical performance of vdW homo-/heterostructures. Based on the bulging test, we realize the first measurement of the interlayer shear stress for bilayer graphene. The interlayer shear effect on the bending rigidity of different 2D materials is quantified, highlighting a breakdown of continuum mechanics models. We further disclose Raman signatures of the interlayer shear mode for chemically doped multilayer graphene and demonstrate the effective modulation of the weak interlayer coupling. We advocate that such a Raman shear mode is expected for other 2D materials and their detection will offer a direct probe of interlayer interactions.

In summary, the graphene-based interfacial mechanical behaviors that are rich in shear deformations are systematically investigated. The results from this thesis will provide deep insights into the mechanical design of graphene-based composite materials and devices, as well as paving the pathway for the new direction in nanomechanical studies of 2D materials and vdW heterostructures.

List of Publications

Journal publications during the Ph.D. period (in chronological order before December 2017)

1. **G. Wang**, Z. Dai, Y. Wang, P. Tan, L. Liu, Z. Xu, Y. Wei, R. Huang, Z. Zhang. Measuring Interlayer Shear Stress in Bilayer Graphene. *Phys. Rev. Lett.*, 2017, 119, 036101.
2. **G. Wang**, E. Gao, Z. Dai, L. Liu, Z. Xu, Z. Zhang. Degradation and Recovery of Graphene/Polymer Interfaces under Cyclic Mechanical Loading. *Compos. Sci. Technol.*, 2017, 149, 220.
3. **G. Wang**, X. Li, Y. Wang, Z. Zheng, L. Liu, Z. Dai, X. Qi, Z. Cheng, Z. Xu, P. Tan, Z. Zhang. Interlayer Coupling Behaviors of Boron Doped Multilayer Graphene. *J. Phys. Chem. C*, 2017, 121, 26034
4. **G. Wang**, L. Liu, Z. Dai, H. Hu, Q. Dai and Z. Zhang. Tuning the Interfacial Mechanical Behaviors of Monolayer Graphene/PMMA Nanocomposites. *ACS Appl. Mater. Interfaces*, 2016, 8, 22554.
5. Z. Dai, **G. Wang**, L. Liu, Y. Wei, Z. Zhang. Mechanical behavior and properties of hydrogen bonded graphene/polymer nano-interfaces. *Compos. Sci. Technol.*, 2016, 136, 1.
6. Y. Zhang, L. Liu, B. Sun, **G. Wang**, Z. Zhang. Preparation of lipophilic graphene oxide derivates via a concise route and its mechanical reinforcement in thermoplastic polyurethane. *Compos. Sci. Technol.*, 2016, 134, 36.
7. **G. Wang**, L. Liu, Z. Dai, Q. Liu, H. Miao and Z. Zhang. Biaxial Compressive Behavior of Embedded Monolayer Graphene inside Flexible Poly (methyl methacrylate) Matrix. *Carbon*, 2015, 86, 69.
8. J. Shang, Y. Chen, Y. Zhou, L. Liu, **G. Wang**, X. Li, J. Kuang, Q. Liu, Z. Dai, H. Miao, L. Zhi, Z. Zhang. Effect of folded and crumpled morphologies of graphene oxide platelets on the mechanical performances of polymer nanocomposites. *Polymer*, 2015, 68, 131.
9. Q. Liu, L. Liu, K. Xie, Y. Meng, H. Wu, **G. Wang**, Z. Dai, Z. Wei, Z. Zhang. Synergistic effect of ar-GO/PANI nanocomposite electrode based air working ionic actuator with a large actuation stroke and long-term durability. *J. Mater. Chem. A*, 2015, 3, 8380.
10. J. Han, H. Zhang, M. Chen, **G. Wang**, Z. Zhang. CNT buckypaper/thermoplastic polyurethane composites with enhanced stiffness, strength and toughness. *Compos. Sci. Technol.*, 2014, 103: 63–71.

More publications can be found in my Google Scholar: https://scholar.google.com/citations?user=2glWF0MAAAAJ&hl=en

Acknowledgements

This dissertation is completed under the guidance of Prof. Xiaoping Wu in the University of Science and Technology of China (USTC) and Prof. Zhong Zhang in National Center for Nanoscience and Technology (NCNST), China. I would like to express my deepest gratitude to my advisors for their invaluable help and patient guidance in research, as well as meticulous care and attention in life. In particular, Prof. Zhang supplied opportunities for accessing to the most advanced technology as well as attending the international academic conferences, which made me grow up rapidly to be an independent researcher. His broad perspective and insightful views in mechanics and materials benefit me a lot.

Much appreciation to Prof. Luqi Liu at NCNST who is the co-supervisor for the thesis. She was instrumental in maintaining my focus and constantly reminding me that identifying the research problem is as important as finding the solution. I owe a lot to her for the thorough guidance in the experiment details and paper writing as well as seeking for wonderful collaborations with top research groups, which was helpful in shaping my research skills.

Special thanks are given to Prof. Yujie Wei in Institute of Mechanics (CAS) who for the first time led me to the palace of science when I was a third-year undergraduate working with him on a summer research program. His rigorous attitude and modest affability stimulated my enthusiasm for scientific research and inspired me to fight my way for my Ph.D. study. I am also thankful to Prof. Hui Zhang, Prof. Jun Zhao, Prof. Hao Jin, Dr. Yun Gao, Dr. Jun Kuang, Dr. Jinhua Han and Dr. Qing Liu in NCNST for the support and help during my early Ph.D. period. Many thanks also to Prof. Hong Miao, Prof. Hengan Wu, Prof. Yong Ni and Prof. Xinglong Gong at USTC for their warm encouragement and advice to my academic career.

Many thanks to the people with whom I have had close collaborations. I was lucky that Dr. Zhaohe Dai was in NCNST when I joined the laboratory. His solid theoretical foundation and constructive suggestions greatly improve my studies to a higher level. I hope more collaborations will come from our friendship in the future. I am also grateful to Prof. Zhiping Xu, Dr. Yanlei Wang and Dr. Enlai Gao from

Tsinghua University and Prof. Yuli Chen, Dr. Fei Pan and Yong Ma from Beihang University for their contributions from MD simulations. It is also my pleasure to learn basic knowledge and operation skills of Raman spectroscopy from Prof. Pingheng Tan, Dr. Xiaoli Li and Dr. Xuelu Liu from Institute of Semiconductor (CAS) and Prof. Yanjun Guo from NCNST, which is beneficial throughout my Ph. D. works. I am grateful to Prof. Zhihai Cheng, Prof. Yanlian Yang, Dr. Xiaoyang Zhu and Dr. Zhiyue Zheng from NCNST for their technical support on the AFM testing and relevant discussions.

It is with beautiful memories that I acknowledge the working atmosphere and the company of a great group of colleagues in NCNST. I would like to thank Xiao Xue, Yan Tian, Feng Duan, Weiwei Li, Abolhassan Imani, Lingyu Zhao, Yayun Liu, Hongdong Wang, Wenfu Cai, Chuanxin Weng, Yufeng Guo, Junchao Liu, Yali Yuan, Ruirui Hu, Yanbing Zhang, Mohammad Owais, Xiaomeng Wang, Yuan Hou, Junkai Xiao, Linglong Kong, Yaqin Chen, Tianjiao Wang, Yuantong Li, Shizhao Yuan, Han Tu and many others, for numerous conversations on all sorts of topics and for making my life so much fun.

It is a pleasure to acknowledge my current academic home, NMML Lab in University of Toronto, for its warm and congenial atmosphere, with special thanks to my supervisor Prof. Tobin Filleter and my colleagues for their help and company.

My most important acknowledgment is to my loving family: my parents, my wife and my daughter, who have filled my life with joy and who mean the world to me.

Toronto, Canada
June 2020

Guorui Wang

Contents

Chapter 1
Introduction

1.1 Graphene

Carbon as one of the most commonly distributed elements in nature constitutes life on earth. The valence orbital hybridization of carbon atoms endows carbon materials with a variety of structures and properties, and roughly underlies the classification scheme for carbon allotropes [1]. In particular, the sp^2-hybridization corresponds to graphite comprising a stack of layers of carbon atoms condensing in honeycomb lattice [2]. By setting aside the interlayer interactions, an atomically thin two-dimensional (2D) crystal is obtained, known as graphene. Interestingly, graphene can be further wrapped into zero-dimensional (0D) fullerene and rolled into one-dimensional (1D) carbon nanotubes (CNT), serving as the basic building block for carbon nanomaterials, as shown in Fig. 1.1.

Since its discovery in 2004, graphene has spawned surging interests and spurred a rapidly increasing number of studies from researchers in various communities. Graphene is now well acknowledged as the "super" material due to a wealth of fascinating properties [4]. Albeit being only one-atom thick, defect-free graphene can be completely impermeable to gases and liquids, owing to its hexagonal structures. Such an impermeability has been demonstrated through a graphene-sealed microchamber with monocrystalline walls and the leakage rate was claimed to reach 10^{-4} atoms/s, enabling the permeation of just a few helium atoms per hour [5]. Given the extremely strong in-plane σ-bond formed between carbon atoms, graphene is expected to exhibit superior mechanical properties. Specifically, the first experimental manifestation via atomic force microscopy (AFM) nanoindentation showed that graphene is the strongest material ever measured, featured with the Young's modulus of ~1 TPa and the fracture strength of ~130 GPa [6]. A recent study by Cui et al. further revealed an ultrahigh fatigue life (>10^9 cycles) of graphene even at high stress levels [7]. These high mechanical performances empower graphene to be the ideal candidate material for the reinforcement of composite [8]. In addition, as a zero-gap semiconductor, the conduction and valence bands of graphene touch at the Brillouin zone corners, as shown in Fig. 1.2 [9]. An unusual linear energy dispersion

G. Wang, *Characterization and Modification of Graphene-Based Interfacial Mechanical Behavior*, Springer Theses, https://doi.org/10.1007/978-981-15-8029-1_1

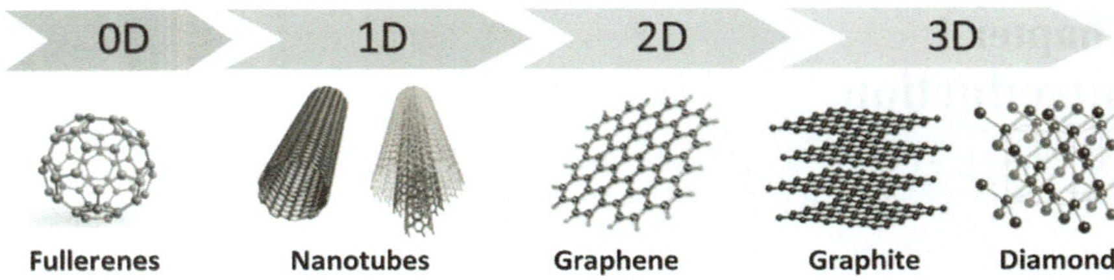

Fig. 1.1 Crystal structure of carbon nanomaterials with different dimensions. Reprinted with permission of Ref. [3], copyright 2014, One Central Press (OCP)

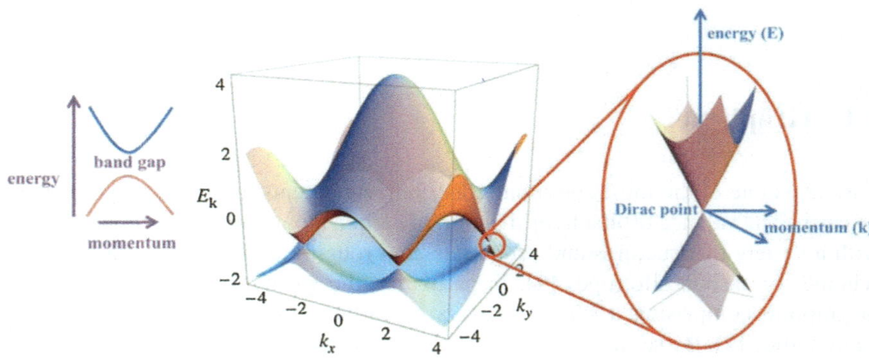

Fig. 1.2 Electronic band structure of the graphene honeycomb lattice. Reprinted with permission of Ref. [9], copyright 2009, American Physical Society

relationship is observed, which makes electrons like the massless relativistic Dirac fermions with a vanishing density of states at the Fermi level. Consequently, the reported carrier mobility in graphene can reach up to 200,000 cm^2/Vs, nearly 140 times that of silicon [10]. Correspondingly, the electrical resistivity of graphene is only around 10^{-6} Ω cm, much lower than that of metals (e.g. copper and silver) [11], rendering graphene to be highest conductive material in the world. The zero-gap band structure also provides graphene with an optical transparency of 97.7% [12], thus granting graphene with huge potential in applications of transparent electrodes, solar cells and flexible electronics [13–15]. More importantly, graphene was demonstrated to exhibit a record high thermal conductivity (~5300 W/mK), almost one order of magnitude higher than that of common metal materials (200–400 W/mK) [16]. This facilitates the heat dissipation and improve the thermal stability of large-scale integrated circuits at the nanoscale [17].

To date, tremendous efforts have been made towards the graphene commercialization from lab to market [18, 19]. The production of graphene in a reliable, controllable, and scalable manner has become the prerequisite to bring us closer to practical applications of graphene [20]. To this end, numerous methods have been developed for the

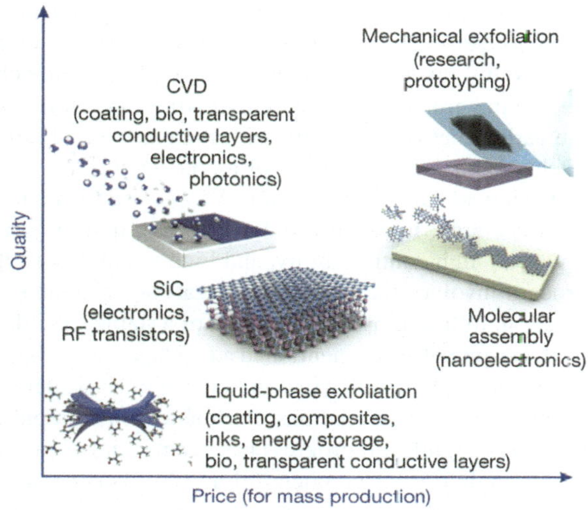

Fig. 1.3 Several methods of mass-production of graphene allow a wide choice in terms of size, quality and price for any particular application. Reprinted with permission of Ref. [4], copyright 2012, Nature Publishing Group

preparation of graphene, such as mechanical exfoliation [21], chemical vapor deposition (CVD) [22, 23], chemical reduction of graphene derivatives [24], liquid-phase exfoliation [25], epitaxial growth [26], and so forth (Fig. 1.3). In particular, Scotch tape-based mechanical exfoliation was developed to give birth to the first graphene flake by Geim's group. The exfoliation mechanism lies in that graphene flakes can be peeled off from highly ordered pyrolytic graphite (HOPG) or natural graphite by external normal forces, based on the weak van der Waals interactions between stacked layers. The thickness of obtained flakes on tapes can be further reduced by repeating the separation procedures until reaching single-atom layers. In principle, the mechanically exfoliated graphene exhibits high-quality, defect-free structures and retain intrinsic properties; however, this method is extremely labor-intensive and time-consuming. The low yield in the production as well as uncontrollable thickness and lateral dimensions limited its application to the laboratory research [27]

The normal exfoliation mechanics can also be extended to liquid-phase exfoliation method that requires dispersion of powdered graphite in a solvent, followed by sonication and centrifugation. This simple method is intriguing and opens a new vista for the large-scale and low-cost production of graphene, yet concerns arise from the extremely low graphene concentration that is far from practical application [25]. The selection of ideal solvents is also critical to the exfoliation efficiency and stability [28]. Alternatively, high-shear mixing method based on the fluid dynamics has been recently proposed to produce defect-free graphene in liquid volumes up to hundreds of liters and beyond [29]. Besides, the shear exfoliation tends to generate thinner flakes compared to the normal exfoliation, due to the superlubric sliding between graphene layers [30].

Similarly, chemical reduction of graphene derivatives also rests on the sonication process, where the graphite is first subjected to the chemical oxidation. During the oxidation, the intercalation of ions between graphene layers leads to the weakening of interlayer interaction and the following sonication allows for isolating graphene oxide (GO) layers from graphite oxide [31]. Nonetheless, even the reduction helps restore the graphene lattice by removing oxygen functional groups, the damages in the carbon networks remain with the presence of structural defects, which greatly affect the physical properties of graphene [32].

Chemical vapor deposition (CVD) is one of the most commonly used approach for large-area, high-uniformity, and high-quality graphene preparation due to its low cost, high controllability, and high scalability [20]. Typically, the CVD growth process involves the activation of gaseous reactants and the subsequent chemical reaction, followed by the formation of a stable solid deposit over a suitable substrate [33]. Despite of abovementioned advantages, a gap between currently available CVD and ideal graphene still exists in terms of intrinsic defective grain boundaries, surface wrinkles and contaminations. Fortunately, with rapid development of CVD growth strategies, significant progress has been achieved in successful fabrication of meter-sized single-crystal graphene films [34], superclean graphene films [35, 36] and ultra-fat, wrinkle-free graphene films [37], which are beneficial to the industrial applications.

1.2 Raman Spectra of Graphene

The atomic thickness and high flexibility of graphene pose considerable challenges in manipulation and characterization with conventional methods. In this scenario, Raman spectroscopy has emerged as a versatile workhorse for the non-invasive characterization of graphene with high resolution [38]. Raman spectroscopy relies upon a light scattering, whereby a molecule scatters incident light from a high-intensity laser light source. Most of the scattered light is at the same frequency as the laser source, known as Rayleigh scattering, which does not provide useful information; however, a small amount of light experiences inelastic scattering and induces the variation in the frequency, which is called Raman scattering. Since Raman scattering depends on the chemical structures of materials, Raman spectroscopy has proven to be highly useful to extract the fingerprint information of graphene such as layer numbers and edges, as well as the effect of strain, doping, defects, temperature and magnetic fields, according to the peak intensity, position and full width at half-maximum (FWHM) [39].

1.2.1 Applications in Structural Characterization

Pristine, defect-free graphene commonly exhibits two characteristic peaks, as identified as Raman G-band at ~1580 cm^{-1} and 2D-band at ~2670 cm^{-1} [40]. The G-band is assigned to a doubly degenerate phonon mode (E_{2g} symmetry) at the Brillouin zone center, reflecting the in-plane vibration of sp^2 carbon atoms, while the 2D-band is originated from a two phonon double resonance Raman process [41].

As shown in Fig. 1.4a, Raman spectra of graphene display a strong dependence on layer numbers [41]. Typically, the number of layers (n) can be identified based on the relative intensity, shape, and position of the G- and 2D-peaks for few-layer graphene ($n < 10$). In particular, the 2D-band of monolayer graphene can be fitted with a sharp and symmetric peak, roughly 4 times higher than the G-band. As graphene thickness increases, it can be seen that the Raman G-band intensity increases almost linearly up to approximately 10 layers, and then decreases for thicker graphite sheets [42]. This enables the visualization of the distribution of graphene flakes with different thickness via the Raman contour mapping of G-band, as shown in the inset of Fig. 1.4b. On the other hand, an asymmetrical broadening accompanied with a blueshift is visible in 2D-band with increasing thickness as presented in Fig. 1.4c [43], which can be explicable by the splitting of the electronic band structure [40]. For instance, Raman 2D-band of a bilayer graphene with AB stacking can be well fitted with four Lorentzians, each with a FWHM of ~24 cm^{-1} [44]. One thing should be noted is

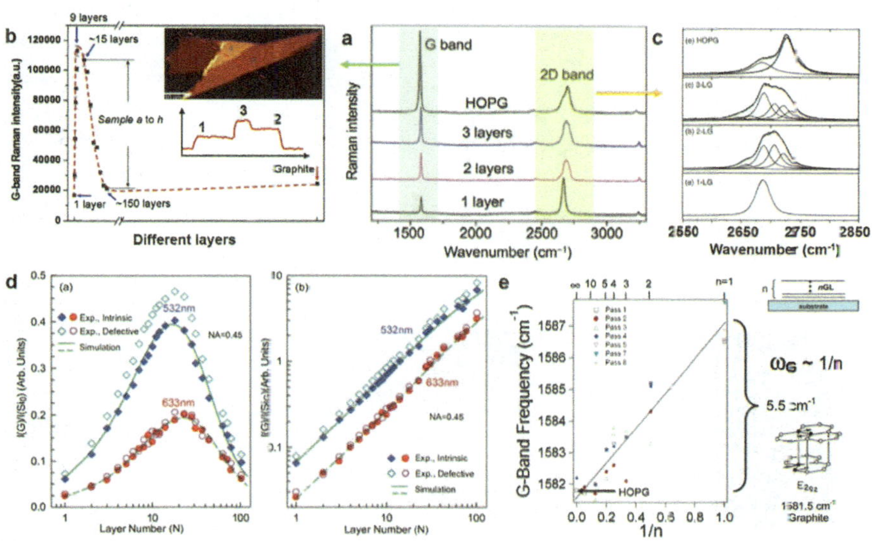

Fig. 1.4 **a** Raman spectra of graphene with different layers as well as that of HOPG. Reprinted with permission of Ref. [41], copyright 2009, Tsinghua University Press. **b** The G-band Raman intensity of graphene sheets as a function of number of layers. Reprinted with permission of Ref. [42], copyright 2008, American Institute of Physics Publishing. The inset shows Raman image plotted by intensity of the G-band and the corresponding the cross-section. Reprinted with permission of Ref. [41], copyright 2009, Tsinghua University Press. **c** The measured 2D-band for graphene with different layers as well as that of HOPG. Reprinted with permission of Ref. [43], copyright 2009, Elsevier. **d** The experimental and theoretical Raman intensity ratio $I(G)/I(Si_0)$ and $I(G)/I(Si_G)$ as a function of layer numbers. Reprinted with permission of Ref. [45], copyright 2015, Royal Society of Chemistry. **e** G-band frequency versus $1/n$, where n is the layer number Reprinted with permission of Ref. [46], copyright 2006, American Chemical Society

that, for more than 5 layers, Raman 2D-band becomes hardly distinguishable from that of bulk graphite. In this case, Tan et al. recently proposed an efficient approach, where the intensity of the Si peak from bare SiO_2/Si substrates $I(Si_0)$ or from SiO_2/Si substrates underneath graphene flakes $I(Si_G)$ is used as a probe for the layer number detection. Such a robust technique is able to identify graphene flakes up to 100 layers for any substrate orientation and laser polarization, as shown in Fig. 1.4d [45]. In addition, it was also evidenced that the G-band exhibited a redshift with increasing layer numbers. An almost linear dependence on $1/n$ was uncovered and described as $\omega_G(n) = \omega_G(\infty) + \beta/n$, thus offering an alternative way to determine the thickness of graphene [46].

Another issue worthy of close scrutiny is the defects or disorders presented in graphene lattice, leading to the appearance of Raman D-band at ~1350 cm^{-1} and the D′-band at ~1620 cm^{-1} in spectrum. Figure 1.5a presents the Raman spectra of defective graphene subjected to the ion bombardment, where a two-stage feature can be clearly observed, recalling the well-established amorphization trajectory proposed by Ferrari et al. to classify disorders in graphite [47, 48]. In detail, the first stage corresponds to the structure transition from single-crystal to nanocrystalline. The defect density (n_D) can be described by the average distance between defects (L_D):

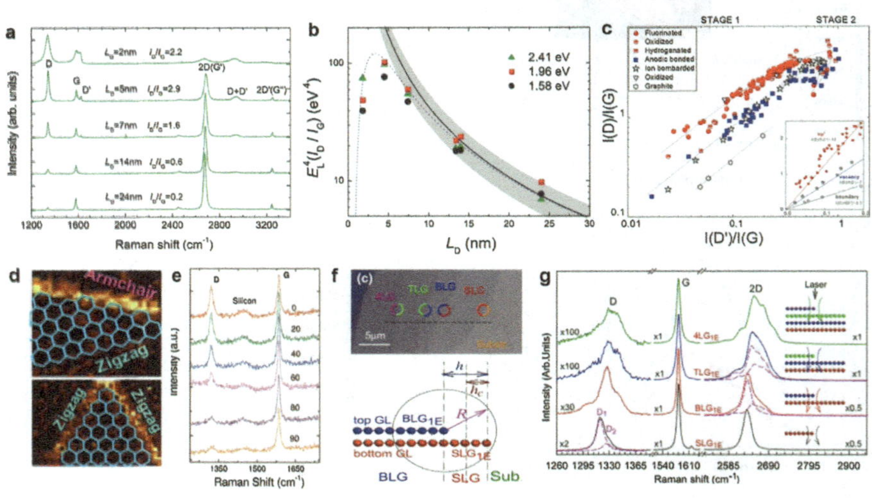

Fig. 1.5 **a** Raman spectra of five ion bombarded monolayer graphene flakes with different L_D. **b** $E_L{}^4$ (I_D/I_G) as a function of L_D. Reprinted with permission of Ref. [47], copyright 2011, American Chemical Society. **c** I_D/I_G as a function of $I_{D'}/I_G$ showing different defect types. Reprinted with permission of Ref. [50], copyright 2012, American Chemical Society. **d** Raman imaging of D-band intensity results from edges with angles 30° and 60°. Reprinted with permission of Ref. [52], copyright 2008, American Institute of Physics Publishing. **e** Raman spectra of the edge measured for different incident polarization. Reprinted with permission of Ref. [53], copyright 2009, American Chemical Society. **f** Optical image and schematic illustration of a laser spot focused at graphene with well-defined edges. **g** Raman spectra of multilayer graphene edge as indicated by the circles in **f**. Reprinted with permission of Ref. [54], copyright 2015, Elsevier

$$n_D\left(\text{cm}^{-2}\right) = 10^{14}/\left(\pi L_D^2\right) \tag{1.1}$$

As shown in Fig. 1.5a, both D- and D′-band show up and stiffen with decreasing L_D and hence increasing n_D. Quantitatively, I_D is proportional to the total number of defects within the Raman laser spot with a size of L_L, i.e. $I_D \propto (L_L/L_D)^2$; I_G is proportional to the total area probed by the laser ($I_D \propto L_D^2$), thus giving rise to $I_D/I_G \propto 1/L_D^2$. The detailed relationship is expressed as (shadowed region in Fig. 1.5b) [39]:

$$L_D^2\left(\text{nm}^2\right) = \frac{4.3 \times 10^3}{E_L^4\left(\text{eV}^4\right)}\left[\frac{I_D}{I_G}\right]^{-1} \tag{1.2}$$

where E_L is the excitation energy. I_D/I_G is found to reach the maximum as L_D decreases to ~3 nm. This actually corresponds to the average distance an electron–hole pair travels before scattering with a phonon, below which contributions of structural defects will not sum independently anymore. When the defect density further increases to make $L_D < 3$ nm, I_D/I_G exhibits a precipitous decrease, suggestive of the amorphization of graphene. In the second stage, sp^2 domains become smaller and hexagonal rings become fewer and more distorted, until they open up [47]. Instead, I_D is proportional to the number of rings in the laser spot, leading to [39]:

$$L_D^2\left(\text{nm}^2\right) = 5.9 \times 10^{-2} E_L^4\left(\text{eV}^4\right)\frac{I_D}{I_G} \tag{1.3}$$

Accordingly, D-band broadens and D′-band gradually merges with G-band, while the well-defined 2D-band becomes almost indiscernable. Therefore, even two different stages could have the same I_D/I_G, we are able to discriminate them by combining the considerable change of D-band FWHM. Such an evolution of defect density can also be fitted by a phenomenological model, which is in conceptual agreement with the amorphization trajectory [49].

In addition to the estimation of defect density, Raman spectroscopy is also capable of investigating the nature of defects, as depicted in Fig. 1.5c. Specifically, $I_D/I_{D'}$ can reach as high as ~13 for defects associated with sp^3 hybridization while drop to ~7 for the vacancy-type defects. The boundary-type defect such as the edge of graphene gives an $I_D/I_{D'}$ of around 3.5. These results are consistent with ab initio calculations, where the hopping defects that describe graphene containing vacancies have $I_D/I_{D'}$ ~ 10.5, and on-site defects in sp^3-defective graphene yield an $I_D/I_{D'}$ ~ 1.3 [50].

Referring to the graphene edges, commonly viewed as extended defects, the orientation of the carbon hexagons with respect to the edge axis in the so-called armchair or zigzag arrangements can be distinguished spectroscopically by the Raman D-band intensity [51]. It was found that the armchair edge shows a more pronounced D-peak compared to the zigzag edge, as shown in Fig. 1.5d, because the double resonance process can only be fulfilled at an armchair edge while forbidden for a zigzag edge [52]. In addition, the Raman D-band at the edge of graphene is also dependent on the incident light polarization, which is maximum for polarization parallel to the edge and minimum when perpendicular (Fig. 1.5e) [53]. For multilayer graphene having individual layers misaligned with adjacent layers, the profile of Raman D-band becomes more complicated convoluting signals from both top and bottom layers. Given the limited Raman laser spot size (R ~ 500 nm), there exists a critical distance (h_c) near the edge and I_D relies on misalignment distance (h). When $h > h_c$, the edge of the top graphene layer laid on the n-layer graphene presents an identical line shape to that of disordered ion-implanted ($n + 1$)-layer graphene; on the contrary, when $h < h_c$, the edge can be considered as an alignment configuration, then I_D will be proportional to the active edge length of graphene layers [54].

1.2.2 Applications in Mechanical Measurement

Any changes in the atomic structure in a crystalline solid when subjected to deformations are reflected in the lattice vibration and hence Raman spectrum. By probing the Raman peaks including the peak shift and width, Raman spectroscopy provides precise measurements of local strains and gains comprehensive insight into the mechanical properties of carbon nanomaterials such as graphene [55].

As graphene is subjected to uniaxial tensile strains, the C=C bond is elongated and induces the variation of phonon vibration frequency. As a result, Raman G-band exhibits an obvious redshifting tendency as shown in Fig. 1.6a. Due to the Poisson's ratio effect, a compression is applied to graphene in the direction perpendicular to the stretching, yielding an asymmetric strain field. The shear component of the strain breaks the crystalline symmetry and causes the splitting of G-band, where the G^- mode is parallel to the applied strain, thus experiencing larger softening, while the G^+ mode perpendicular to it. The split two peaks share similar FWHM of 12 cm^{-1} that is roughly constant as a function of strain. By linear fitting of the Raman shift as a function of strain, the Raman shift rate can be deduced as $\partial\omega_{G^+}/\partial\varepsilon = -10.8$ and $\partial\omega_{G^-}/\partial\varepsilon = -31.7$ cm^{-1}/%. In contrast, the Raman 2D-band displays a higher sensitivity to the applied strain and its Raman shift rate is almost doubled that of G-band, $\partial\omega_{2D}/\partial\varepsilon = -64$ cm^{-1}/% [56]. The strain-induced splitting of 2D-band was also reported at a higher strain level, stemming from the anisotropic modifications of the phonon dispersion together with changes in electronic structures. Raman shift rates for each split Raman 2D-peak is strongly dependent on the strain direction, i.e. armchair or zigzag (Fig. 1.6b) [57]. Likewise, if graphene suffers from a compressive strain, the Raman peaks will blueshift accompanied with a peak splitting [58, 59]. When the graphene is under a biaxial strain, the shear deformation term will be canceled so that the G-peak does not split.

In addition to evaluating the strain level by Raman shift, the FWHM of Raman peaks can also be considered as a measure of nanometer-scale strain variations in graphene. Neumann et al. found that Raman G-band width was significantly increased in regions where the graphene flake features bubbles and folds (Fig. 1.6c). By using a magnetic field of 8 T to constrain the influence of the electronic contributions on the Raman G-band width, the structure deformation induced non-uniform strains was

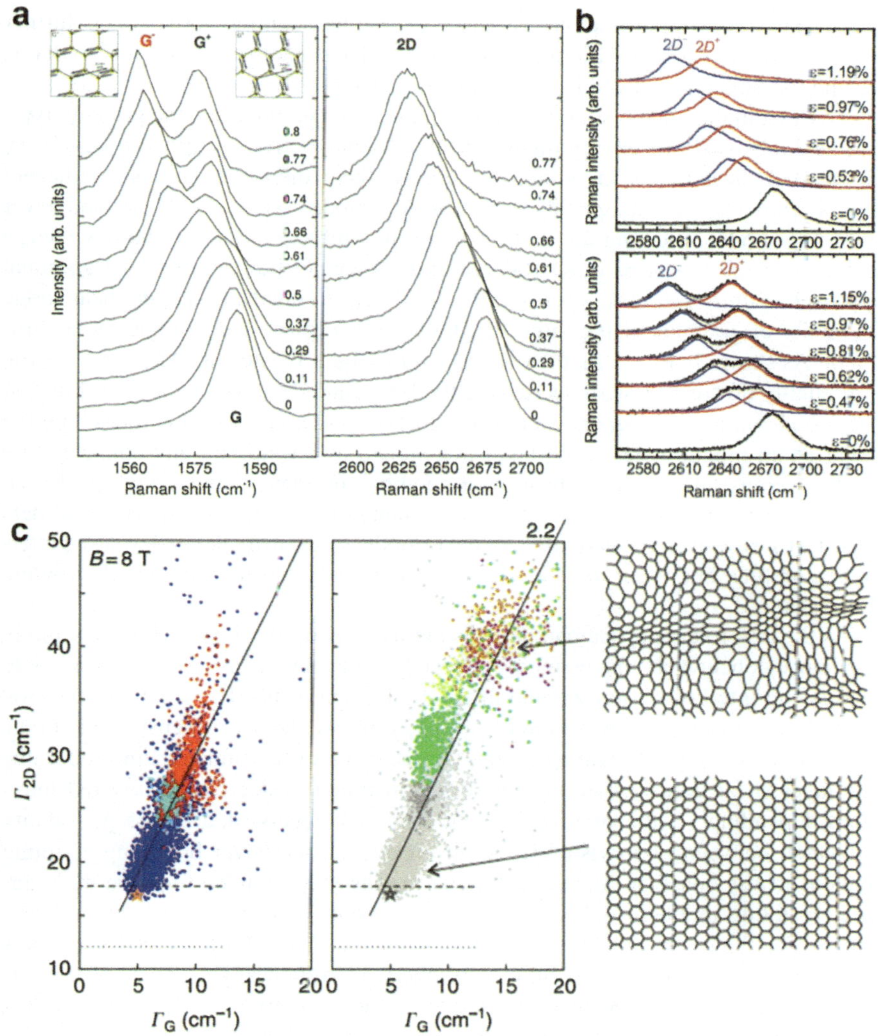

Fig. 1.6 **a** Raman G- and 2D-bands of monolayer graphene under uniaxial strain. Reprinted with permission of Ref. [56], copyright 2009, American Physical Society. **b** Evolution of the 2D-band as a function of strain along the armchair and the zigzag directions, respectively. Reprinted with permission of Ref. [57], copyright 2011, American Physical Society. **c** The width of Raman 2D-band as a function of the width of Raman G-band recorded from four different graphene flakes on hexagonal boron nitride (hBN) and silicon substrate, respectively. Reprinted with permission of Ref. [60], copyright 2015, Nature Publishing Group

believed to mainly contribute to the observed line broadenings [60]. This is further evidenced by the slope of Raman 2D-band to G-band width ratio (Γ_{2D}/Γ_G) that is around 2.2 when dominated by the strain effect [61].

Apart from the widely studied high-frequency range, in reality, the graphite typically exhibits two low-energy mode: one is a Raman-active E_{2g} mode (~42 cm^{-1}), also called shear mode, involving the relative shear motion of atoms in adjacent planes; the other is an optically inactive B_{2g} mode (~126 cm^{-1}), named layer breathing mode, reflecting the motions in normal direction. Due to the low energy and weak peak intensity, these Raman features cannot be captured by traditional approaches until Tan et al. reported the first measurement of Raman shear mode (C-mode) of graphene using three BragGrate notch filters and a single monochromator. As shown in Fig. 1.7a, the C-band of graphite is located at ~43.5 cm^{-1}, while it gradually downshifts with decreasing thickness and drop to 31 cm^{-1} for bilayers [62]. Such a thickness dependence of peak positions can be well interpreted by the linear chain model, where each layer is assumed as a rigid ball and connected by elastic springs taking into only nearest-neighbor interlayer interactions [63]. The interlayer coupling strength can then be estimated by fitting the experimental data (Fig. 1.7b). However, the hardening of the C mode is not attributed to the varied interlayer coupling, but rather to an increase of the overall restoring force with growing thickness [62].

As the low-frequency Raman mode is highly sensitive to the interlayer coupling, it is hence conceivable that changing the spacing or relative layer orientation would modify the interlayer coupling and alters the Raman peaks. For example, twisting multiple graphene layers was found to greatly improve the peak intensity of both the shear mode and layer breathing mode. On the basis of the linear chain model, the interlayer shear coupling strength was demonstrated to decrease at twisted interfaces depending on the relative orientation of the layers, due to the periodicity mismatch between two twisted layers; however, the interlayer breathing coupling remains nearly constant. Such a technique can also be extended to probe the interlayer coupling modes of other 2D layered materials, like transition metal chalcogenides that have stronger electron–phonon coupling [64]. Furthermore, some new peaks tend to appear in twist bilayer graphene at certain angles, such as *R*-mode corresponding to 9° and *R'*-mode related to 12° as seen in Fig. 1.7c, d, which offers a way to identify the twisting angles between graphene layers [65].

1.3 Current Research Status of Graphene-Based Interfacial Mechanics

1.3.1 Some Important Issues in This Field

Intermolecular forces embrace all forms of the matter. These forces result from the electromagnetic interactions between the electrons and nuclei forming molecules

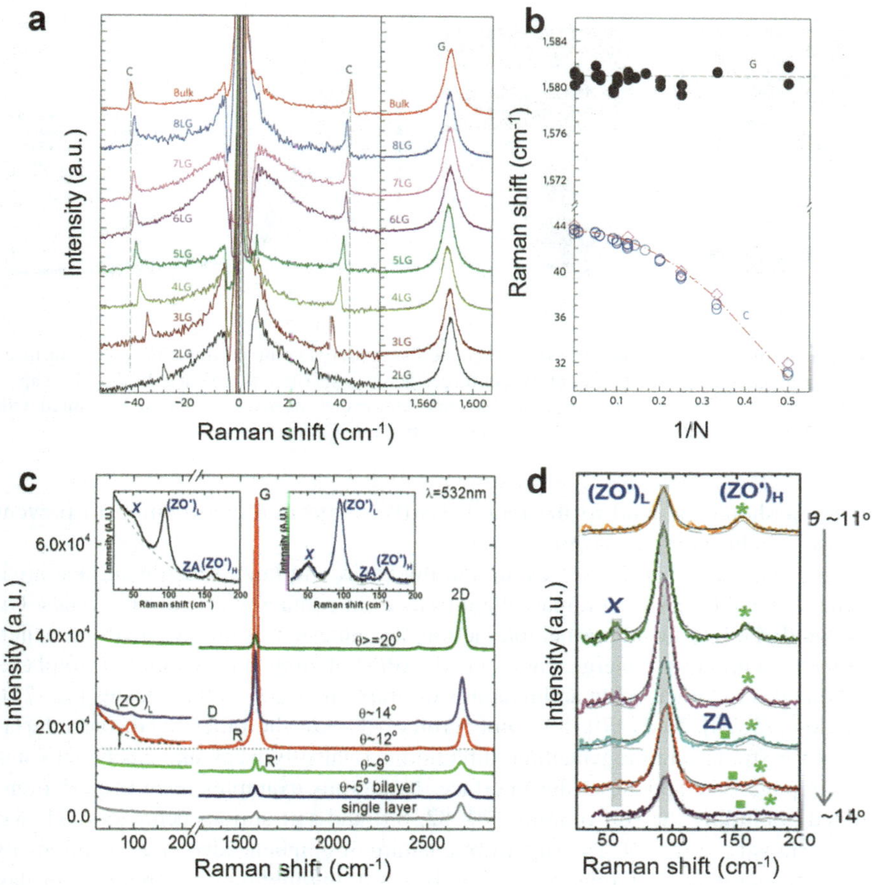

Fig. 1.7 **a** Raman spectra for the C-peak spectral region (left) and G-peak spectral region (right). **b** Peak positions of Raman G-band and C-band as a function of inverse layer number. Reprinted with permission of Ref. [62], copyright 2012, Nature Publishing Group. **c** Raman spectra from twisted bilayer graphene domains with different twisting angle θ. The left inset displays a zoomed-in low-energy spectrum. The right inset displays the same spectrum after subtraction of the background envelope. **d** Background-subtracted low-energy Raman spectra from six different bilayer domains with twisting angles between 11° and 14°. Reprinted with permission of Ref. [65], copyright 2013, American Chemical Society

[66]. At the macroscale, these forces are too weak to impact the mechanical behavior of materials; however, as the dimensions shrink to the microscale or even nanoscale, the negligible surface forces start to play an important role in the system response [67, 68]. As an example, the large surface-to-volume ratio of microelectromechanical systems (MEMS) requires van der Waals (vdW) forces to be considered in the device performance, which is a function of roughness (Fig. 1.8). With size and separation of these surfaces further downscale in nanoelectromechanical systems (NEMS), more

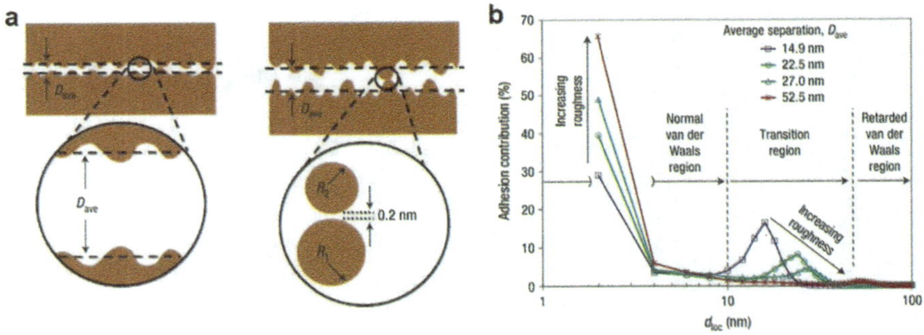

Fig. 1.8 **a** The adhesion is dominated by forces across non-contacting portions for smooth surfaces while dominated by normal van der Waals forces across asperities that nearly bridge the gap at rough surfaces. **b** A histogram of the adhesion contributions versus pixel separation. Reprinted with permission of Ref. [69], copyright 2005, Nature Publishing Group

attention should be paid to the retarded vdW interaction contribution to prevent adhesion-induced failures [69].

Likewise, graphene possesses atomic thickness and huge specific surface area, allowing vdW forces to dominate the interfaces in contact with other materials. On one hand, the interfacial normal interaction, i.e. adhesion, is relatively strong which competes with elastic energies associated with stretching and bending of graphene and dictates the microscale architectures of graphene, such as the buckles [72–74], bubbles [75–78], folds [79, 80], and scrolls [81]. On the other hand, the interfacial shear strength in the tangential direction is relatively weak and easily activates the slippage of graphene at the interface. Numerous examples have been demonstrated, including the self-rotation [70, 82, 83] and self-sliding behaviors [71, 84, 85] as shown in Fig. 1.9. The superlubric nature of graphene also greatly softens its elastic response to stretching or bending [86–88], highlighting a complex interplay

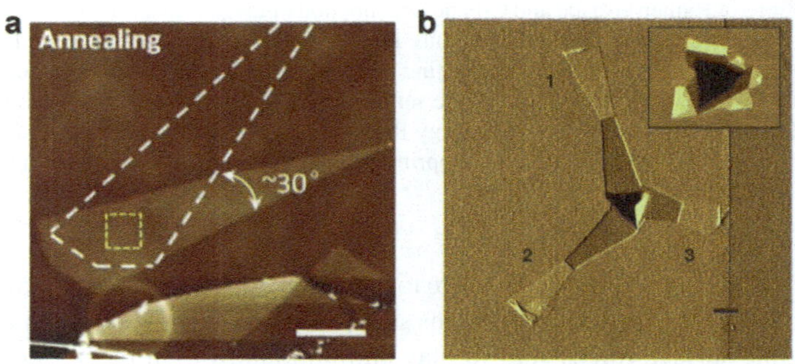

Fig. 1.9 **a** Thermal-induced rotation of graphene on hBN. Reprinted with permission of Ref. [70], copyright 2016, American Physical Society. **b** Self-growth of graphene ribbons nucleated from single fretted nanoindentation contact Reprinted with permission of Ref. [71], copyright 2016, Nature Publishing Group

between measurements of intrinsic elasticity and interfacial property. To this end, graphene can serve as a versatile platform for the interfacial mechanical studies at the nanoscale, to understand and leverage such interesting phenomena [89–91]. From the perspective of fundamental science, quantifying the interfacial interactions is particularly instrumental for modeling the deformation and assembly of microstructures made of graphene.

From the application point-of-view, the extraordinary physical properties of graphene lend well to a wide range of applications, such as flexible electronics [92], MEMS/NEMS [93], and nanocomposites [94], where interfacial adhesion and friction are important factors governing the performance and reliability of the devices and materials. Specifically, in the application of flexible electronics, the interfacial strength determines the maximum strain that graphene can sustain. A strong interface can enable an efficient stress transfer and a high elastic strain limit, which is advantageous to the flexibility and stretchability. Unfortunately, the weakness of vdW-dominated interface usually initiates interfacial failures such as cracking, slipping and delamination, especially under complex loading conditions (e.g. stretching, compressing, bending, twisting) in the practical use [95–98]. These failures are likely to result in degradation and even ultimate damage of the devices (Fig. 1.10a–c).

The interfacial adhesion also plays an important role in graphene-based nanomechanical devices. For example, manufacturing of the device firstly requires graphene

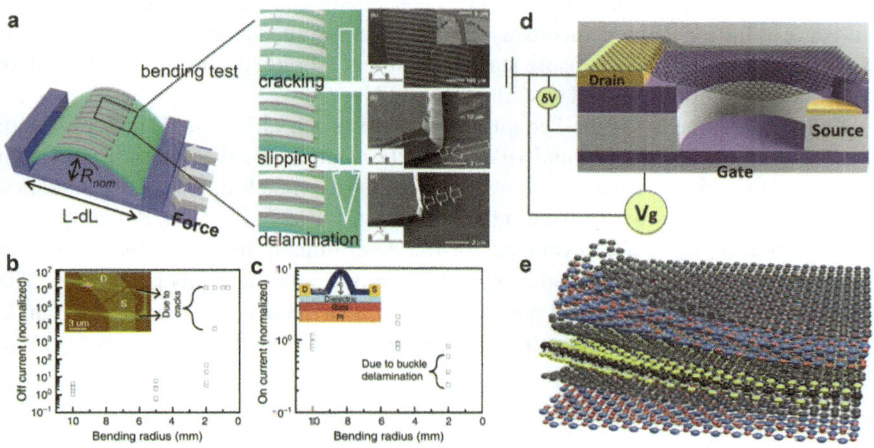

Fig. 1.10 **a** Schematic illustration of various failure mechanisms (i.e., cracking, slipping, or delamination) at sufficiently large degrees of bending. Reprinted with permission of Ref. [95], copyright 2008, Wiley. **b** Mechanical studies of flexible field effect transistor (FET), where the exponential increase in off current is due to cracks in the gate dielectric shown in the inset. **c** The on current degrades owing to the occurrence of buckling delamination. Reprinted with permission of Ref. [96], copyright 2014, Nature Publishing Group. **d** 3D schematic of a 3-terminal graphene NEMS switch. Reprinted with permission of Ref. [101], copyright 2014, Wiley. **e** Van der Waals heterostructures formed by integrating the 2D nanosheets. Reprinted with permission of Ref. [113], copyright 2016, Nature Publishing Group

films to be grown and transferred to suitable substrates. The engineering of the peeling, stamping, and other fabrication procedures rests on the adhesion of the graphene to both the host and target substrates [99]. Thus, a thorough knowledge of the adhesion energies between graphene and different substrate materials is critical. In addition, the nanomechanical devices such as graphene switches can be fabricated by simply depositing graphene over the predefined trenches or microwells by making use of the ultrastrong adhesion of graphene membranes (Fig. 1.10d) [100, 101]. However, such strong adhesive forces sometimes result in irreversible stiction problems that is detrimental to device performance [102]. It is hence imperative to engineer the graphene substrate interface to address the adhesion challenges in NEMS devices.

Classic views of conventional composites imply that the mechanical reinforcement is strongly dependent on the interfacial strength between the filler and surrounding matrix [103–106]. The mechanism lies in that the stress transfer across the interfaces enables the fillers to bear most of the loads. As the use of graphene as the nanofiller will introduce more interface areas, the effect of interfacial interactions on the overall performance of nanocomposite is expected to be more significant. Numerous studies have manifested that the failures happened in graphene-based nanocomposite prefers taking the form of interfacial slippage/pullout rather than fractures in the fiber-reinforced composites [107, 108]. More importantly, vdW forces readily reform after the bond breaking during the deformation, which maintains the shear stress transfer at the interface. The consequently enhanced energy dissipation can serve as the mechanical mechanism for toughening and damping augmentation of nanocomposite, helping to ensure stability and low vibratory loads in a wide variety of structural components and systems [109–111].

Graphene research has sparked immense interests in a variety of 2D materials with diverse electronic properties and vdW interactions allow for the integration of such nanoscale materials without the constraints of lattice matching, to create so-called vdW heterostructures with intriguing physics (Fig. 1.10e) [112–114]. Recently, vdW pick-up method has been developed for the deterministic transfer which relies on the competition of adhesive forces between the different materials [115, 116]. The tunable interlayer coupling via adjusting stacking sequence and the twisting angle between layers will open opportunities for new physics and applications for vdW heterostructures [82, 117, 118].

1.3.2 Nanoscale Mechanical Characterization Methods

Exploration of the interfacial mechanical properties and underlying atomic mechanism not only helps us to gain a better understanding of 2D materials, but also acts as a key step towards developing new applications. Nevertheless, the atomic thickness and low-dimensionality of graphene disable traditional characterization techniques and pose vexing challenges in the nondestructive sample preparation, high-precision manipulation and positioning, and high-resolution mechanical measurement.

To date, significant progress has been achieved in characterizing the interfacial behavior of graphene. Ni et al. was the first to report the stretching of graphene on a polymer substrate and reveal stress transfer mechanism via in situ micro-Raman technique [119]. Later, Gong et al. observed the graphene relaxation at relatively high strain during the tensile deformation, in terms of the irregular band shifts as seen in Fig. 1.11a, which is ascribed to the onset of interfacial slippage [120]. Mapping the local strain along the graphene in a polymer matrix allows the interfacial shear strength to be evaluated based on the shear-lag model (Fig. 1.11b). Following this method, the Galiotis's group further systematically investigated the effect of wrinkles, edge geometry, lateral size and thickness on the interfacial shear properties as well as failure modes of the graphene/polymer interface [121–124]. To better interpret the interfacial slip phenomenon, Jiang et al. adopted a nonlinear shear-lag model and estimated the interfacial shear strength based on the strain gradient within the sliding regions initiated from the edge of graphene flakes. Meanwhile, the occurrence of buckle delamination was identified as an additional failure mechanism during the unloading process, which enables the measurement of interfacial adhesion energy between graphene and polymer substrate based on a fracture mechanics analysis [73].

In addition to the buckling-based method, double cantilever beam (DCB) fracture mechanics testing and blister test offer alternate ways to measure the adhesion properties between graphene and other substrates, like silicon and metal. As an example, in DCB fracture experiments, the CVD grown graphene was glued to a target substrate with epoxy and a force is applied to delaminate the graphene from the copper foil, as illustrated in Fig. 1.11c. If the strain energy release rate exceeds the adhesion energy of the interfaces in the specimen, crack starts to grow and propagates through the interfaces. The measured load–displacement responses associated with delamination along the interfaces were then utilized to determine the adhesion energy, which featured a separation rate dependence [125, 126]. Bunch's group first extended the micro-scale blister test to mechanically exfoliated graphene suspended over circular cavities. By increasing the pressure difference across the graphene membrane, microbubbles grew stably until the delamination at the interface. Due to the impermeability of graphene, the gas leakage rate is low enough for reliable measurements of bubble profiles by AFM, based on which the adhesion energy can be deduced as shown in Fig. 1.11d [100].

One thing should be noted that multilayer graphene is commonly regarded as a thin plate assuming the interlayer shear resistance to be infinite and forbid the relative sliding between layers. However, the reality is the opposite: Zheng et al. designed self-retraction experiments on square graphite mesas and observed the microscale superlubricity at the incommensurate interface between graphene layers [85, 127]. Besides, the interfacial shear strength as well as the cleavage energy or interlayer binding energy can be measured simultaneously, as indicated in Fig. 1.11e [128]. Such an incommensurate state can also be determined by Raman spectroscopy, as the turbostratic stacking corresponds to a broad symmetric 2D-peak as seen in Fig. 1.11f. Frank et al. looked into more details regarding evolution of the Raman 2D-band of bilayer graphene. The discrepancy in Raman shift rates of four sub-bands

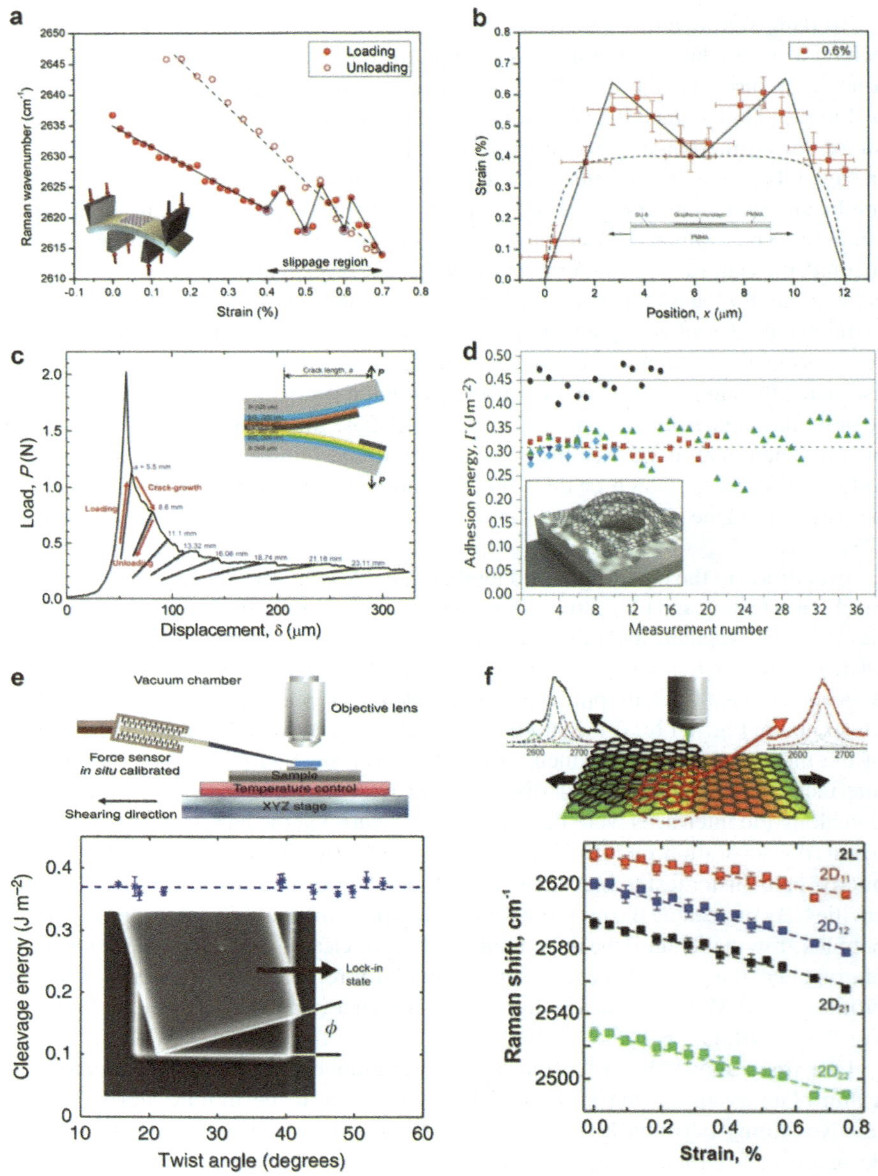
a
Loading
Unloading
Raman wavenumber (cm^{-1})
slippage region
Strain (%)
b
0.6%
Strain (%)
Position, x (μm)
c
Crack length, a
P
Load, P (N)
Loading
Crack-growth
Unloading
Displacement, δ (μm)
d
Adhesion energy, Γ (J m^{-2})
Measurement number
e
Vacuum chamber
Objective lens
Force sensor
in situ calibrated
Sample
Temperature control
XYZ stage
Shearing direction
Cleavage energy (J m^{-2})
Lock-in state
φ
Twist angle (degrees)
f
2L
$2D_{11}$
$2D_{12}$
$2D_{21}$
$2D_{22}$
Raman shift, cm^{-1}
Strain, %

◀**Fig. 1.11** **a** Shift of the 2D-band peak position for monolayer graphene composite. Reprinted with permission of Ref. [120], copyright 2010, Wiley. The inset depicts the schematic of four-point bending test. Reprinted with permission of Ref. [56], copyright 2009, American Physical Society. **b** Variation of graphene strain in the direction of the tensile axis (x) across a single monolayer at 0.6% matrix strain. Reprinted with permission of Ref. [120], copyright 2010, Wiley. **c** Adhesion energy measurements on CVD grown graphene on copper measured by double cantilever beam fracture tests. Reprinted with permission of Ref. [126], copyright 2012, American Chemical Society. **d** Adhesion energies for graphene membranes containing different layers measured by blister test. Reprinted with permission of Ref. [100], copyright 2011, Nature Publishing Group. The inset depicts the schematic of graphene bubble device. Reprinted with permission of Ref. [130], copyright 2011, Nature Publishing Group. **e** The cleavage energy as a function of twist angle measured through shearing the lower graphite flake relative to the upper one in the superlubric state. Reprinted with permission of Ref. [128], copyright 2015, Nature Publishing Group. **f** The transition of Raman 2D-band feature of bilayer graphene from Bernal to turbostratic stacking during the tensile deformation, with four subpeaks showing different Raman shift rates Reprinted with permission of Ref. [129], copyright 2012, American Chemical Society

suggest a different stress transfer efficiency across two different interfaces, namely, graphene/graphene interface and graphene/polymer interface [129].

1.4 Outline

To address the interfacial mechanics issues and technical challenges, the thesis aims to develop universal micro-/nano-mechanical experimental platforms and extend continuum mechanics models for quantifying key mechanical parameters as well as interpreting nanoscale deformation and failure mechanisms at graphene-based interfaces.

In Chap. 2, we focus on the interfacial issues present in graphene/polymer nanocomposite. Two interfacial failure mechanisms are identified as shear sliding during stretching and buckling under compression. The degradation mechanism of the interface is revealed when subjected to a cyclic loading, while the chemical functionalization strategy is proposed for interfacial strengthening. In detail, impacts of interface bonding types (e.g. vdW and hydrogen bonds) on both key interfacial parameters (e.g. interfacial shear strength and adhesion) and failure modes of the nanocomposite were clarified for the first time at the microscopic level. This is followed by a discussion on the optimal design of high-performance graphene-based nanocomposite via interfacial engineering.

In Chap. 3, special attention is paid to the interfacial stability problems in graphene-based flexible electronics applications. The thermal effect on the mechanical response of graphene embedded in polymers is systematically investigated, as classified by three different stages in terms of microstructural characteristics and stress states. The interface stability and recoverability are examined, and a critical failure strain is evaluated based on the mechanical instability theory.

In Chap. 4, we combine Raman spectroscopy and AFM technique to offer a comprehensive mechanical measurement of graphene bubbles. We highlight a previously neglected shear slippage of graphene on the silicon substrate, in challenge with the clamped boundary assumption widely applied in application of nanomechanical devices. Taking into the slippery boundary, a modified Hencky's solution is exploited to extract the interfacial shear resistance at graphene/SiO_2 interfaces. The interplay between the interfacial friction and adhesion was explored, which determines the geometric characteristics of stable delamination.

In Chap. 5, we concentrate on the vdW interfaces between 2D material layers. Following the experiments and theoretical modeling in Chap. 4, the interlayer slip between graphene layers is captured, allowing the interlayer shear stress to be measured. We discuss the interlayer shear effect on the elastic response of multilayer 2D materials, accounting for the breakdown of continuum mechanics model in prediction of bending rigidity. Chemical doping is further performed to tune the interlayer coupling, as evidenced by Raman shear mode measurements.

In Chap. 6, we summarize our findings and present the novelties in the thesis. We also discuss a number of avenues for further research.

References

1. Su DS, Perathoner S, Centi G (2013) Nanocarbons for the development of advanced catalysts. Chem Rev 113(8):5782–5816
2. Belenkov EA, Greshnyakov VA (2013) Classification of structural modifications of carbon. Phys Solid State 55(8):1754–1764
3. Kanoun O, Müller C, Benchirouf A, Sanli A, Bouhamed A, Al-Hamry A et al (2014) Potential of flexible carbon nanotube films for high performance strain and pressure sensors. In: Nanotechnology for optics and sensors. One Central (OCP)
4. Novoselov KS, Fal V, Colombo L, Gellert P, Schwab M, Kim K (2012) A roadmap for graphene. Nature 490(7419):192–200
5. Sun PZ, Yang Q, Kuang WJ, Stebunov YV, Xiong WQ, Yu J et al (2020) Limits on gas impermeability of graphene. Nature 579(7798):229–232
6. Lee C, Wei X, Kysar JW, Hone J (2008) Measurement of the elastic properties and intrinsic strength of monolayer graphene. Science 321(5887):385–388
7. Cui T, Mukherjee S, Sudeep PM, Colas G, Najafi F, Tam J et al (2020) Fatigue of graphene. Nat Mater 19(4):405–411
8. Stankovich S, Dikin DA, Dommett GH, Kohlhaas KM, Zimney EJ, Stach EA et al (2006) Graphene-based composite materials. Nature 442(7100):282–286
9. Castro Neto AH, Guinea F, Peres NMR, Novoselov KS, Geim AK (2009) The electronic properties of graphene. Rev Mod Phys 81(1):109–162
10. Bolotin KI, Sikes KJ, Jiang Z, Klima M, Fudenberg G, Hone J et al (2008) Ultrahigh electron mobility in suspended graphene. Solid State Commun 146(9–10):351–355
11. Geim AK (2009) Graphene: status and prospects. Science 324(5934):1530–1534
12. Nair RR, Blake P, Grigorenko AN, Novoselov KS, Booth TJ, Stauber T et al (2008) Fine structure constant defines visual transparency of graphene. Science 320(5881):1308–1308
13. Bae S, Kim H, Lee Y, Xu X, Park J-S, Zheng Y et al (2010) Roll-to-roll production of 30-inch graphene films for transparent electrodes. Nat Nanotechnol 5(8):574–578
14. Li X, Fan L, Li Z, Wang K, Zhong M, Wei J et al (2012) Boron doping of graphene for graphene-silicon p-n junction solar cells. Adv Energy Mater 2(4):425–429

15. Eda G, Fanchini G, Chhowalla M (2008) Large-area ultrathin films of reduced graphene oxide as a transparent and flexible electronic material. Nat Nanotechnol 3(5):270–274
16. Balandin AA, Ghosh S, Bao WZ, Calizo I, Teweldebrhan D, Miao F et al (2008) Superior thermal conductivity of single-layer graphene. Nano Lett 8(3):902–907
17. Lin Y-M, Valdes-Garcia A, Han S-J, Farmer DB, Meric I, Sun Y et al (2011) Wafer-scale graphene integrated circuit. Science 332(6035):1294–1297
18. Ferrari AC, Bonaccorso F, Fal'ko V, Novoselov KS, Roche S, Boggild P et al (2015) Science and technology roadmap for graphene, related two-dimensional crystals, and hybrid systems. Nanoscale 7(11):4598–4810
19. Kong W, Kum H, Bae SH, Shim J, Kim H, Kong L et al (2019) Path towards graphene commercialization from lab to market. Nat Nanotechnol 14(10):927–938
20. Lin L, Deng B, Sun J, Peng H, Liu Z (2018) Bridging the gap between reality and ideal in chemical vapor deposition growth of graphene. Chem Rev 118(18):9281–9343
21. Geim AK, Novoselov KS (2007) The rise of graphene. Nat Mater 6(3):183–191
22. Li XS, Cai WW, An JH, Kim S, Nah J, Yang DX et al (2009) Large-area synthesis of high-quality and uniform graphene films on copper foils. Science 324(5932):1312–1314
23. Yan K, Fu L, Peng H, Liu Z (2013) Designed CVD growth of graphene via process engineering. Acc Chem Res 46(10):2263–2274
24. Moon IK, Lee J, Ruoff RS, Lee H (2010) Reduced graphene oxide by chemical graphitization. Nat Commun 1(1):1–6
25. Hernandez Y, Nicolosi V, Lotya M, Blighe FM, Sun Z, De S et al (2008) High-yield production of graphene by liquid-phase exfoliation of graphite. Nat Nanotechnol 3(9):563–568
26. Usachov DY, Fedorov AV, Petukhov AE, Vilkov OY, Rybkin AG, Otrokov MM et al (2015) Epitaxial B-graphene: large-scale growth and atomic structure. ACS Nano 9(7):7314–7322
27. Yi M, Shen Z (2015) A review on mechanical exfoliation for the scalable production of graphene. J Mater Chem A 3(22):11700–11715
28. Shen J, He Y, Wu J, Gao C, Keyshar K, Zhang X et al (2015) Liquid phase exfoliation of two-dimensional materials by directly probing and matching surface tension components. Nano Lett 15(8):5449–5454
29. Paton KR, Varrla E, Backes C, Smith RJ, Khan U, O'Neill A et al (2014) Scalable production of large quantities of defect-free few-layer graphene by shear exfoliation in liquids. Nat Mater 13(6):624–630
30. Rokni H, Lu W (2018) Nanoscale probing of interaction in atomically thin layered materials. ACS Cent Sci 4(2):288–297
31. Ang PK, Wang S, Bao Q, Thong JT, Loh KP (2009) High-throughput synthesis of graphene by intercalation–exfoliation of graphite oxide and study of ionic screening in graphene transistor. ACS Nano 3(11):3587–3594
32. Pei S, Cheng H-M (2012) The reduction of graphene oxide. Carbon 50(9):3210–3228
33. Zhang Y, Zhang L, Zhou C (2013) Review of chemical vapor deposition of graphene and related applications. Acc Chem Res 46(10):2329–2339
34. Xu X, Zhang Z, Dong J, Yi D, Niu J, Wu M et al (2017) Ultrafast epitaxial growth of metre-sized single-crystal graphene on industrial Cu foil. Sci Bull 62(15):1074–1080
35. Zhang J, Jia K, Lin L, Zhao W, Quang HT, Sun L et al (2019) Large-area synthesis of superclean graphene via selective etching of amorphous carbon with carbon dioxide. Angew Chem Int Ed Engl 58(41):14446–14451
36. Sun L, Lin L, Wang Z, Rui D, Yu Z, Zhang J et al (2019) A force-engineered lint roller for superclean graphene. Adv Mater 31(43):1902978
37. Yuan G, Lin D, Wang Y, Huang X, Chen W, Xie X et al (2020) Proton-assisted growth of ultra-flat graphene films. Nature 577(7789):204–208
38. Wu JB, Lin ML, Cong X, Liu HN, Tan PH (2018) Raman spectroscopy of graphene-based materials and its applications in related devices. Chem Soc Rev 47(5):1822–1873
39. Ferrari AC, Basko DM (2013) Raman spectroscopy as a versatile tool for studying the properties of graphene. Nat Nanotechnol 8(4):235–246

40. Ferrari A, Meyer J, Scardaci V, Casiraghi C, Lazzeri M, Mauri F et al (2006) Raman spectrum of graphene and graphene layers. Phys Rev Lett 97(18):187401
41. Ni Z, Wang Y, Yu T, Shen Z (2008) Raman spectroscopy and imaging of graphene. Nano Res 1(4):273–291
42. Wang YY, Ni ZH, Shen ZX, Wang HM, Wu YH (2008) Interference enhancement of Raman signal of graphene. Appl Phys Lett 92(4):043121
43. Malard L, Pimenta M, Dresselhaus G, Dresselhaus M (2009) Raman spectroscopy in graphene. Phys Rep 473(5):51–87
44. Malard L, Nilsson J, Elias D, Brant J, Plentz F, Alves E et al (2007) Probing the electronic structure of bilayer graphene by Raman scattering. Phys Rev B 76(20):201401
45. Li XL, Qiao XF, Han WP, Lu Y, Tan QH, Liu XL et al (2015) Layer number identification of intrinsic and defective multilayered graphenes up to 100 layers by the Raman mode intensity from substrates. Nanoscale 7:8135–8141
46. Gupta A, Chen G, Joshi P, Tadigadapa S, Eklund P (2006) Raman scattering from high-frequency phonons in supported n-graphene layer films. Nano Lett 6(12):2667–2673
47. Cancado LG, Jorio A, Ferreira EH, Stavale F, Achete CA, Capaz RB et al (2011) Quantifying defects in graphene via Raman spectroscopy at different excitation energies. Nano Lett 11(8):3190–3196
48. Ferrari AC (2007) Raman spectroscopy of graphene and graphite: disorder, electron–phonon coupling, doping and nonadiabatic effects. Solid State Commun 143(1–2):47–57
49. Lucchese MM, Stavale F, Ferreira EHM, Vilani C, Moutinho MVO, Capaz RB et al (2010) Quantifying ion-induced defects and Raman relaxation length in graphene. Carbon 48(5):1592–1597
50. Eckmann A, Felten A, Mishchenko A, Britnell L, Krupke R, Novoselov KS et al (2012) Probing the nature of defects in graphene by Raman spectroscopy. Nano Lett 12(8):3925–3930
51. Cancado LG, Pimenta MA, Neves BR, Dantas MS, Jorio A (2004) Influence of the atomic structure on the Raman spectra of graphite edges. Phys Rev Lett 93(24):247401
52. You YM, Ni ZH, Yu T, Shen ZX (2008) Edge chirality determination of graphene by Raman spectroscopy. Appl Phys Lett 93(16):163112
53. Casiraghi C, Hartschuh A, Qian H, Piscanec S, Georgi C, Fasoli A et al (2009) Raman spectroscopy of graphene edges. Nano Lett 9(4):1433–1441
54. Li QQ, Zhang X, Han WP, Lu Y, Shi W, Wu JB et al (2015) Raman spectroscopy at the edges of multilayer graphene. Carbon 85:221–224
55. Gao Y, Li L, Tan P, Liu L, Zhang Z (2010) Application of Raman spectroscopy in carbon nanotube-based polymer composites. Chin Sci Bull 55(35):3978–3988
56. Mohiuddin T, Lombardo A, Nair R, Bonetti A, Savini G, Jalil R et al (2009) Uniaxial strain in graphene by Raman spectroscopy: G peak splitting, Grüneisen parameters, and sample orientation. Phys Rev B 79(20):205433
57. Yoon D, Son YW, Cheong H (2011) Strain-dependent splitting of the double-resonance Raman scattering band in graphene. Phys Rev Lett 106(15)
58. Tsoukleri G, Parthenios J, Papagelis K, Jalil R, Ferrari AC, Geim AK et al (2009) Subjecting a graphene monolayer to tension and compression. Small 5(21):2397–2402
59. Frank O, Tsoukleri G, Parthenios J, Papagelis K, Riaz I, Jalil R et al (2010) Compression behavior of single-layer graphenes. ACS Nano 4(6):3131–3138
60. Neumann C, Reichardt S, Venezuela P, Drogeler M, Banszerus L, Schmitz M et al (2015) Raman spectroscopy as probe of nanometre-scale strain variations in graphene. Nat Commun 6:8429
61. Lee JE, Ahn G, Shim J, Lee YS, Ryu S (2012) Optical separation of mechanical strain from charge doping in graphene. Nat Commun 3:1024
62. Tan PH, Han WP, Zhao WJ, Wu ZH, Chang K, Wang H et al (2012) The shear mode of multilayer graphene. Nat Mater 11(4):294–300
63. Zhang X, Qiao XF, Shi W, Wu JB, Jiang DS, Tan PH (2015) Phonon and Raman scattering of two-dimensional transition metal dichalcogenides from monolayer, multilayer to bulk material. Chem Soc Rev 44(9):2757–2785

64. Wu JB, Hu ZX, Zhang X, Han WP, Lu Y, Shi W et al (2015) Interface coupling in twisted multilayer graphene by resonant Raman spectroscopy of layer breathing modes. ACS Nano 9(7):7440–7449
65. He R, Chung TF, Delaney C, Keiser C, Jauregui LA, Shand PM et al (2013) Observation of low energy Raman modes in twisted bilayer graphene. Nano Lett 13(8):3594–3601
66. Park Soojin SM-K (2011) Interface science and composites, 1st edn. Elsevier, Academic Press, Amsterdam, Boston
67. Maboudian R (1997) Critical review: adhesion in surface micromechanical structures. J Vac Sci Technol B 15(1):1
68. Israelachvili JN (2011) Intermolecular and surface forces. Academic Press
69. Delrio FW, de Boer MP, Knapp JA, David Reedy E Jr, Clews PJ, Dunn ML (2005) The role of van der Waals forces in adhesion of micromachined surfaces. Nat Mater 4(8):629–634
70. Wang D, Chen G, Li C, Cheng M, Yang W, Wu S et al (2016) Thermally induced graphene rotation on hexagonal boron nitride. Phys Rev Lett 116(12):126101
71. Annett J, Cross GL (2016) Self-assembly of graphene ribbons by spontaneous self-tearing and peeling from a substrate. Nature 535(7611):271–275
72. Zhang K, Arroyo M (2013) Adhesion and friction control localized folding in supported graphene. J Appl Phys 113(19):193501
73. Jiang T, Huang R, Zhu Y (2014) Interfacial sliding and buckling of monolayer graphene on a stretchable substrate. Adv Funct Mater 24(3):396–402
74. Deng S, Berry V (2016) Wrinkled, rippled and crumpled graphene: an overview of formation mechanism, electronic properties, and applications. Mater Today 19(4):197–212
75. Boddeti NG, Koenig SP, Long R, Xiao J, Bunch JS, Dunn ML (2013) Mechanics of adhered, pressurized graphene blisters. J Appl Mech 80(4):040909
76. Wang P, Liechti KM, Huang R (2016) Snap transitions of pressurized graphene blisters. J Appl Mech 83(7):071002
77. Khestanova E, Guinea F, Fumagalli L, Geim AK, Grigorieva IV (2016) Universal shape and pressure inside bubbles appearing in van der Waals heterostructures. Nat Commun 7:12587–12596
78. Dai Z, Hou Y, Sanchez DA, Wang G, Brennan CJ, Zhang Z et al (2018) Interface-governed deformation of nanobubbles and nanotents formed by two-dimensional materials. Phys Rev Lett 121(26):266101
79. Zhao J, Deng Q, Ly TH, Han GH, Sandeep G, Rummeli MH (2015) Two-dimensional membrane as elastic shell with proof on the folds revealed by three-dimensional atomic mapping. Nat Commun 6:8935–8940
80. Cranford S, Sen D, Buehler MJ (2009) Meso-origami: folding multilayer graphene sheets. Appl Phys Lett 95(12):123121
81. Xu Z, Buehler MJ (2010) Geometry controls conformation of graphene sheets: membranes, ribbons, and scrolls. ACS Nano 4(7):3869–3876
82. Ribeiro-Palau R, Zhang CJ, Watanabe K, Taniguchi T, Hone J, Dean CR (2018) Twistable electronics with dynamically rotatable heterostructures. Science 361(6403):690–693
83. Feng X, Kwon S, Park JY, Salmeron M (2013) Superlubric sliding of graphene nanoflakes on graphene. ACS Nano 7(2):1718–1724
84. Zheng Q, Jiang B, Liu S, Weng Y, Lu L, Xue Q et al (2008) Self-retracting motion of graphite microflakes. Phys Rev Lett 100(6):067205
85. Liu Z, Yang J, Grey F, Liu JZ, Liu Y, Wang Y et al (2012) Observation of microscale superlubricity in graphite. Phys Rev Lett 108(20):205503
86. Wei X, Meng Z, Ruiz L, Xia W, Lee C, Kysar JW et al (2016) Recoverable slippage mechanism in multilayer graphene leads to repeatable energy dissipation. ACS Nano 10(2):1820–1828
87. Pan F, Wang G, Liu L, Chen Y, Zhang Z, Shi X (2019) Bending induced interlayer shearing, rippling and kink buckling of multilayered graphene sheets. J Mech Phys Solids 122:340–363
88. Liu Y, Xu Z, Zheng Q (2011) The interlayer shear effect on graphene multilayer resonators. J Mech Phys Solids 59(8):1613–1622

89. Lee C, Li Q, Kalb W, Liu XZ, Berger H, Carpick RW et al (2010) Frictional characteristics of atomically thin sheets. Science 328(5974):76–80
90. Koren E, Lörtscher E, Rawlings C, Knoll AW, Duerig U (2015) Adhesion and friction in mesoscopic graphite contacts. Science 348(6235):679–683
91. Deng Z, Smolyanitsky A, Li QY, Feng XQ, Cannara RJ (2012) Adhesion-dependent negative friction coefficient on chemically modified graphite at the nanoscale. Nat Mater 11(12):1032–1037
92. You R, Liu YQ, Hao YL, Han DD, Zhang YL, You Z (2019) Laser fabrication of graphene-based flexible electronics. Adv Mater 32(15):1901981
93. Bunch JS, van der Zande AM, Verbridge SS, Frank IW, Tanenbaum DM, Parpia JM et al (2007) Electromechanical resonators from graphene sheets. Science 315(5811):490–493
94. Kinloch IA, Suhr J, Lou J, Young RJ, Ajayan PM (2018) Composites with carbon nanotubes and graphene: an outlook. Science 362(6414):547–553
95. Park SI, Ahn JH, Feng X, Wang SD, Huang YG, Rogers JA (2008) Theoretical and experimental studies of bending of inorganic electronic materials on plastic substrates. Adv Funct Mater 18(18):2673–2684
96. Akinwande D, Petrone N, Hone J (2014) Two-dimensional flexible nanoelectronics. Nat Commun 5:5678–5689
97. Chen M, Wang Z, Ge X, Wang Z, Fujisawa K, Xia J et al (2019) Controlled fragmentation of single-atom-thick polycrystalline graphene. Matter 2(3):666–679
98. Na SR, Wang X, Piner RD, Huang R, Willson CG, Liechti KM (2016) Cracking of polycrystalline graphene on copper under tension. ACS Nano 10(10):9616–9625
99. Li D, Windl W, Padture NP (2009) Toward site-specific stamping of graphene. Adv Mater 21(12):1243–1246
100. Koenig SP, Boddeti NG, Dunn ML, Bunch JS (2011) Ultrastrong adhesion of graphene membranes. Nat Nanotechnol 6(9):543–546
101. Liu X, Suk JW, Boddeti NG, Cantley L, Wang L, Gray JM et al (2014) Large arrays and properties of 3-terminal graphene nanoelectromechanical switches. Adv Mater 26(10):1571–1576
102. Shi Z, Lu H, Zhang L, Yang R, Wang Y, Liu D et al (2012) Studies of graphene-based nanoelectromechanical switches. Nano Res 5(2):82–87
103. Piggott MR (2002) Load bearing fibre composites. Springer Science & Business Media
104. Hull D, Clyne T (1996) An introduction to composite materials. Cambridge University Press
105. Chen Y, Dai Z, Weng C, Wang G, Liu X, Cong X et al (2018) Engineering the interface in mechanically responsive graphene-based films. RSC Adv 8(63):36257–36263
106. Wan S, Chen Y, Wang Y, Li G, Wang G, Liu L et al (2019) Ultrastrong graphene films via long-chain π-bridging. Matter 1(2):389–401
107. Xu P, Loomis J, Bradshaw RD, Panchapakesan B (2012) Load transfer and mechanical properties of chemically reduced graphene reinforcements in polymer composites. Nanotechnology 23(50):505713
108. Wagner HD, Vaia RA (2004) Nanocomposites: issues at the interface. Mater Today 7(11):38–42
109. Najafi F, Wang G, Mukherjee S, Cui T, Filleter T, Singh CV (2020) Toughening of graphene-based polymer nanocomposites via tuning chemical functionalization. Compos Sci Technol 194:108140
110. Gardea F, Glaz B, Riddick J, Lagoudas DC, Naraghi M (2015) Energy dissipation due to interfacial slip in nanocomposites reinforced with aligned carbon nanotubes. ACS Appl Mater Interfaces 7(18):9725–9735
111. Suhr J, Koratkar N, Keblinski P, Ajayan P (2005) Viscoelasticity in carbon nanotube composites. Nat Mater 4(2):134–137
112. Geim AK, Grigorieva IV (2013) Van der Waals heterostructures. Nature 499(7459):419–425
113. Liu Y, Weiss NO, Duan X, Cheng H-C, Huang Y, Duan X (2016) Van der Waals heterostructures and devices. Nat Rev Mater 1(9):16042

114. Novoselov KS, Mishchenko A, Carvalho A, Castro Neto AH (2016) 2D materials and van der Waals heterostructures. Science 353(6298):aac9439
115. Fan S, Vu QA, Tran MD, Adhikari S, Lee YH (2020) Transfer assembly for two-dimensional van der Waals heterostructures. 2D Mater 7(2):022005
116. Frisenda R, Navarro-Moratalla E, Gant P, Perez De Lara D, Jarillo-Herrero P, Gorbachev RV et al (2018) Recent progress in the assembly of nanodevices and van der Waals heterostructures by deterministic placement of 2D materials. Chem Soc Rev 47(1):53–68
117. Yankowitz M, Chen S, Polshyn H, Zhang Y, Watanabe K, Taniguchi T et al (2019) Tuning superconductivity in twisted bilayer graphene. Science 363(6431):1059–1064
118. Cao Y, Fatemi V, Fang S, Watanabe K, Taniguchi T, Kaxiras E et al (2018) Unconventional superconductivity in magic-angle graphene superlattices. Nature 556(7699):43–50
119. Ni ZH, Yu T, Lu YH, Wang YY, Feng YP, Shen ZX (2008) Uniaxial strain on graphene: Raman spectroscopy study and band-gap opening. ACS Nano 2(11):2301–2305
120. Gong L, Kinloch IA, Young RJ, Riaz I, Jalil R, Novoselov KS (2010) Interfacial stress transfer in a graphene monolayer nanocomposite. Adv Mater 22(24):2694–2697
121. Li Z, Kinloch IA, Young RJ, Novoselov KS, Anagnostopoulos G, Parthenios J et al (2015) Deformation of wrinkled graphene. ACS Nano 9(4):3917–3925
122. Androulidakis C, Koukaras EN, Rahova J, Sampathkumar K, Parthenios J, Papagelis K et al (2017) Wrinkled few-layer graphene as highly efficient load bearer. ACS Appl Mater Interfaces 9(31):26593–26601
123. Androulidakis C, Sourlantzis D, Koukaras EN, Manikas AC, Galiotis C (2019) Stress-transfer from polymer substrates to monolayer and few-layer graphenes. Nanoscale Adv 1(12):4972–4980
124. Manikas AC, Pastore Carbone MG, Woods CR, Wang Y, Souli I, Anagnostopoulos G et al (2019) Stress transfer at the nanoscale on graphene ribbons of regular geometry. Nanoscale 11(30):14354–14361
125. Na SR, Suk JW, Tao L, Akinwande D, Ruoff RS, Huang R et al (2015) Selective mechanical transfer of graphene from seed copper foil using rate effects. ACS Nano 9(2):1325–1335
126. Yoon T, Shin WC, Kim TY, Mun JH, Kim TS, Cho BJ (2012) Direct measurement of adhesion energy of monolayer graphene as-grown on copper and its application to renewable transfer process. Nano Lett 12(3):1448–1452
127. Yang J, Liu Z, Grey F, Xu Z, Li X, Liu Y et al (2013) Observation of high-speed microscale superlubricity in graphite. Phys Rev Lett 110(25):255504
128. Wang W, Dai S, Li X, Yang J, Srolovitz DJ, Zheng Q (2015) Measurement of the cleavage energy of graphite. Nat Commun 6:7853–7859
129. Frank O, Bousa M, Riaz I, Jalil R, Novoselov KS, Tsoukleri G et al (2012) Phonon and structural changes in deformed Bernal stacked bilayer graphene. Nano Lett 12(2):687–693
130. Huang R (2011) Show of adhesive strength. Nat Nanotechnol 6(9):537–538

Chapter 2
Mechanical Behavior at Graphene/Polymer Interfaces Under Uniaxial Tension

2.1 Introduction

Graphene with extremely high stiffness and strength, large surface area, high aspect ratio, low mass density, has been envisaged to be the ideal reinforcement for polymer nanocomposites [1–7]. From classic view of fiber/matrix system, one of the most critical issues that determine the overall performance of composites lies in the interfacial properties, which should be substantial to undergo sufficient stress transfer caused by the stiffness mismatch during deformation [8–10]. Particularly, apparent degradation in interfacial shear stress was observed for fiber reinforced composites after experiencing tens of loading cycles, which impairs the durability and reliability of the composite material [11–13]. This issue tends to be much more notable in graphene-based nanocomposites, where there are as least several orders of magnitude more interfacial area created with the same volume fraction, and the high stiffness of graphene makes such mismatch more distinctive. Therefore, efforts to examine the mechanical response and failure modes as well as quantify interfacial properties are critical and indispensable for these nanoscale interfaces.

Considering two-dimensional nature of graphene, daunting challenges remain in mechanical characterization at the nanoscale. Limited by the tiny size of nanofiller itself, the micromechanical test methods in conventional fiber-based composites (e.g. fiber pullout, fragmentation, and micro-debonding test) cannot be directly employed to measure the interfacial properties in graphene-based composites [14, 15]. Owing to the stress-sensitive characteristics of specific Raman peaks of nanostructured carbon materials, in situ Raman spectroscopy has become a powerful tool to monitor the interfacial shear stress transfer, strain distribution and deformation modes at a microscopic level [16–19]. For example, Gong et al. have monitored the stress transfer in monolayer graphene sandwiched between the PMMA and the SU-8 polymer layer using a four-point bending fixture. The derived interfacial shear strength was on the order of 0.3–0.8 MPa [20]. In fact, once the interfacial shear strength is reached, the relative sliding starts at the interface where the stress is transferred through the friction. Different from what happens at the macroscale, friction and spontaneous

G. Wang, *Characterization and Modification of Graphene-Based Interfacial Mechanical Behavior*, Springer Theses, https://doi.org/10.1007/978-981-15-8029-1_2

adhesion take place even without an applied normal pressure at the nanoscale [21]. Assuming a constant friction that maintains as interfacial shear strength beyond the elastic stage, Jiang et al. developed a nonlinear shear-lag model and measured interfacial frictional stress to be 0.46–0.69 MPa for monolayer graphene-polyethylene terephthalate (PET) matrix nanocomposites [22].

In order to improve the mechanical performance of graphene nanocomposites, more endeavors were made toward the optimal design of interface properties [8, 23–26], and the high specific surface area of graphene provides a rich platform for surface chemistry. Among various types of interfacial bonds (such as covalent bonds, coordination bonds, ionic bonds, hydrogen bonds, and van der Waals forces), building hydrogen bonds to tailor interfaces has garnered much attention especially in biomimetic materials, such as artificial nacre [27, 28]. Although the hydrogen bonds are not that strong as the covalent bonds, they are able to reform after breaking under deformation, resulting in a high frictional stress while ensuring interface slip [29, 30]. In addition, due to the stronger interaction than van der Waals (vdW) forces, the hydrogen bonding is expected to simultaneously improve the strength and toughness of graphene composites [31–33]. However, the underlying reinforcing mechanism as well as impact of interfacial bonding types on the interfacial properties and failure modes of graphene/polymer interface has not been deeply understood yet at the microscopic level, which will be the focus of this chapter.

Here we combined in situ tensile Raman spectroscopy with atomic force microscopy (AFM) to monitor the interfacial stress transfer process and reveal the possible deformation/failure modes of graphene/PMMA interface when subjected to cyclic loading. A nonlinear shear-lag model was adopted and modified to quantify the interfacial properties, such as stress transfer efficiency, interfacial stiffness, shear strength and adhesion energy. Furthermore, the chemical strategy was exploited to engineer interfacial crosslinks beyond vdW interactions, and the impact of surface functionalization and functionalization degree of graphene on the interfacial interaction of nanocomposites was elucidated, which clarifies the importance of optimal design of high-performance graphene-based nanocomposites [34, 35].

2.2 Experimental Methods

2.2.1 Wet Transfer of Graphene

PMMA powder was dissolved in anisole followed with the magnetically stir for 15 min to prepare PMMA solution (2 wt %). A thin layer of PMMA was then spin-coated on the substrate with graphene flake on top. In the case of spin speed of 3000 rpm for 1 min, the film thickness can reach 100 nm. Afterwards the detachment of the PMMA/graphene layer from the initial surface was done by partially etching the surface of SiO_2 with a 1 M NaOH aqueous solution heated at 50 °C. As a result, a PMMA membrane with all of the graphite/graphene sheets attached to it was obtained

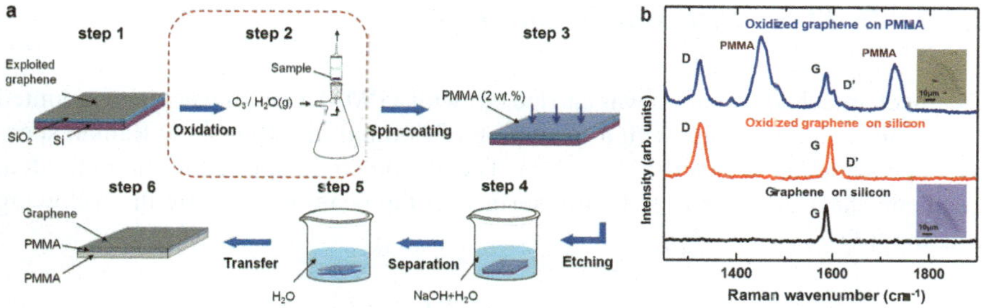

Fig. 2.1 **a** Schematic illustration of the sample preparation process. For pristine PMMA/graphene nanocomposite systems, step 2 was skipped. **b** Raman spectra of pristine graphene sheet deposited onto silicon and oxidized graphene deposited onto silicon and PMMA substrates. The insets show optical images of the graphene sheet on silicon and PMMA substrates

and put in deionized water to wash repeatedly. After drying in a vacuum oven at 60 °C for 12 h, the PMMA capping layer can be removed using acetone (or acetone steam). The flow chart is shown in Fig. 2.1a.

2.2.2 *Oxidation of Graphene*

Oxidization of individual graphene sheet was performed by using a O_3/H_2O gaseous mixture as an oxidizing agent [36]. An exfoliated graphene sheet deposited onto silicon wafer was placed into a homemade reactor, in which O_3 (5 wt % in O_3/O_2 mixture) was continuously passed through the reactor chamber at room temperature during the oxidation process. The oxygen gas flow rate was kept at 150 L/h, and the humidity inside the reactor was kept around 60% as monitored by a hygrometer. The reaction time was varied from 0.5 to 7.0 h to control the functionalization degree of graphene sheets.

Microscopic Fourier transform infrared spectrometer (FTIR) (Reflection mode) (ThermoFisher, Nicolet iN10) was employed to identify the various oxygen-containing groups attached on the graphene sheet. The IR signal was collected for 1 min for each spectrum, followed with subtraction of background signal. The spectra range was set as 675–4000 cm^{-1} with a spectral resolution of $\sim$4 cm^{-1}. The Raman spectroscopy was also used to characterize the defects in graphene with a Renishaw Raman spectrometer using the 514.5 nm line of an Ar laser (Fig. 2.1a). The laser power was kept below 1.0 mW on the sample to avoid laser-induced local heating of the sample. The incident light was polarized along the strain direction. The focused laser spot was approximately 1 μm in diameter, and the spectral resolution was $\sim$1 cm^{-1}.

2.2.3 In-Plane Uniaxial Tensile Test

The mechanical deformation was carried out with a PMMA cantilever beam mounted onto a piezoelectric stage with a resolution of 600 nm in a Renishaw Raman spectroscope setup, as depicted in Fig. 2.2a. The tensile strain applied to the individual graphene sheet deposited on the top surface of the beam is given by the following equation [37]:

$$\varepsilon\left(x, \frac{t}{2}\right) = \frac{3t\delta(L - x)}{2L^3} \tag{2.1}$$

where x is the location in graphene sheet measured from the fixed end of beam, t is the thickness of the beam, L is the span of the beam and δ is the deflection of loading point at the other end of beam, as shown in Fig. 2.2b. To make the assumptions for Eq. (2.1) valid, experimentally, the aspect ratio of span at maximum deflection was ensured to be higher than 10, and the in-plane strain in graphene was within the range from -1.5 to $+1.5\%$ [38]. Remarkably, bending of the cantilever beam would lead to a strain gradient on the top surface, but herein, the strain variation across the graphene flake (along the x direction) was actually less than 0.005% at a given strain level of 1%, that is attributed to the limited span of graphene flake (~10 μm). Consequently, tensile strain within the flake could be regarded as a constant for the estimation of the strain level.

The collection of Raman spectra during loading were carried out in two different ways: on one hand, the Raman laser spot was focused on the center of the graphene flake as the strain increased with an incremental step of 0.05%. The shift of Raman characteristic peaks as a function of strain was obtained; on the other hand, at a given strain level, the Raman laser scanned from one edge to another along the stretching direction with the step of 600 nm. The distribution of the Raman peak position in the graphene sheet can be extracted. To convert the nominal applied strains to actual graphene strains, the measured Raman shift rate was compared to the mean universal value, which for the 2D-band had been found to be 60 cm^{-1}/%, through the formula $\varepsilon\ (\%) = \Delta\omega/|60|$ [39].

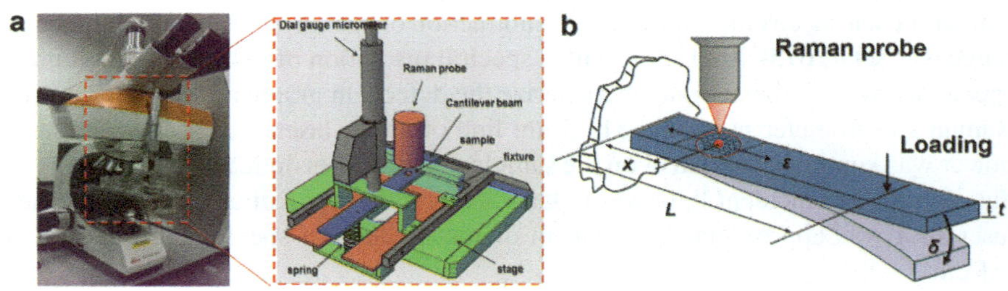

Fig. 2.2 **a** Raman spectroscopy combined with homemade cantilever beam-based loading device. **b** The schematic of cantilever beam as well as corresponding parameters

The surface morphology of graphene was characterized by atomic force microscope (AFM, Dimension 3100) in tapping mode with a scanning frequency of 1 Hz. The RTESP silicon tip was used the experiment, having a resonance frequency of 300 kHz and a cantilever spring constant of 40 N/m.

2.3 Uniaxial Stretching of Graphene on Polymer Substrate

2.3.1 Nonlinear Shear-Lag Model

The shear-lag model was first proposed by Cox et al. and applied to the traditional fiber-reinforced composites [40]. Then Young's group creatively extended it to graphene/polymer systems to interpret the load transfer across interfaces [20]. The basic principle lies in that, when the polymer matrix is subjected to tensile loading, the stress can be transferred from the matrix to the graphene through the interfacial shear stress. Take a flake element as shown in Fig. 2.3, the shear stress at the interface (τ) should be balanced by the force due to the variation of axial stress in the flake ($d\sigma$):

$$-td\sigma = \tau dx \tag{2.2}$$

where t is the graphene thickness and σ is the axial stress along the stretching direction (x). Assuming a perfect bonding at the interface with a constant effective surface stiffness (k), then the relationship holds:

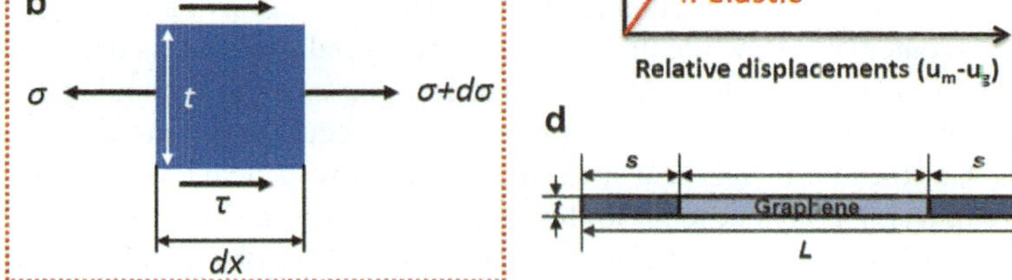

Fig. 2.3 **a** Schematic diagram of a section through the PMMA/graphene system. **b** Force analysis on the element of length from graphene embedded in polymer matrix. **c** Nonlinear shear-lag model. **d** The schematic of graphene flake consisting of two sliding zones at the edges and the elastic zone in between

$$\tau = k(u_m - u_g) \tag{2.3}$$

where u_m and u_g denote the displacement in the matrix and graphene, respectively. Substituting this equation into Eq. (2.2) and differentiating with respect to x lead to:

$$-t\frac{d^2\sigma}{dx^2} = k\left(\frac{du_m}{dx} - \frac{du_g}{dx}\right) \tag{2.4}$$

Given the strain low range during the loading, the linear elasticity assumption is satisfied, so that:

$$\begin{cases} \sigma = E_g\varepsilon_g = E_g\frac{du_g}{dx} \\ \varepsilon_m = \frac{du_m}{dx} \end{cases} \tag{2.5}$$

where E_g is the Young's modulus of graphene, ε_g is the local strain in the graphene sheet, ε_m is the applied strain of bulk matrix. Combining Eqs. (2.4) and (2.5), the following differential equation is obtained:

$$\frac{d^2\varepsilon_g}{dx^2} - \beta^2\varepsilon_g + \beta^2\varepsilon_m = 0 \tag{2.6}$$

$\beta = \sqrt{k/E_g t}$ is the shear-lag parameter which is typically treated as an effective measure for the efficiency of interfacial stress transfer [22]. Considering the boundary condition that the axial stress σ must be zero at two ends of the graphene flake ($\varepsilon_g|_{x=-L/2} = \varepsilon_g|_{x=L/2} = 0$), the distribution of the axial strain and interfacial shear stress along the graphene flake are expressed as:

$$\varepsilon_g = \varepsilon_m\left[1 - \frac{\cosh(\beta x)}{\cosh(\beta L/2)}\right] \tag{2.7}$$

$$\tau = \beta E_g t\varepsilon_m\frac{\sinh(\beta x)}{\cosh(\beta L/2)} \tag{2.8}$$

By Eqs. (2.7) and (2.8), it was found that the maximum strain of graphene occurred at the flake center ($x = 0$) while the maximum interfacial shear stress occurred at the edges ($x = \pm L/2$). As graphene features strong in-plane covalent bonding and weak out-of-plane vdW interactions, failure preferentially occurs at the interface under the loading. Once the interfacial shear stress increases up to a critical level, i.e. interfacial shear strength (τ_c), graphene tends to slip against the substrate. More importantly, due to the breaking-and-reforming nature of vdW forces, the graphene/polymer interface keeps its capacity of transferring loads due to frictional forces, unlike the debonding failure for fiber/polymer interface in conventional interfacial model (Fig. 2.3c). In this case, the interface is divided into two sliding zones emerging from the edges and a non-sliding zone in between, as depicted in Fig. 2.3d. In the sliding zone, the interfacial shear stress is identical to the τ_c, we substitute $\tau = \tau_c$ into Eq. (2.2) and obtains:

$$\tau_c = E_g t \frac{d\varepsilon_g}{dx} \tag{2.9}$$

While Eq. (2.6) is still applicable to the non-sliding zone, the boundary condition requires the axial force in graphene to be continuous at $x = \pm(L/2 - s)$, where s is the length of sliding zone. The strain distribution in the non-sliding zone is

$$\varepsilon_g = \varepsilon_m - \frac{\tau_c}{\beta E_g t} \frac{\cosh(\beta x)}{\sinh[\beta(L/2 - s)]} \tag{2.10}$$

and the sliding zone size can be estimated by

$$\coth\left[\beta\left(\frac{L}{2} - s\right)\right] + \beta s = \frac{\beta E_g t \varepsilon_m}{\tau_c} \tag{2.11}$$

As the strain continuously increases, the sliding zones gradually propagate toward the center until the entire flake slips.

2.3.2 *In Situ Raman Measurement*

As indicated above, Raman spectroscopy has been successfully employed to measure the in-plane strain in graphene due to the high sensitivity of Raman shifts to follow the evolution of strain [41–44]. Figure 2.4a presents the evolution of the Raman 2D band recorded at the center of monolayer graphene flake, in response to the bending deformation. The two-stage features are clearly exhibited, as further summarized in Fig. 2.4b. During the bending deformation, the Raman 2D band initially displays a linear downshift up to 0.7% with a rate of -49 cm^{-1}/%, and then steps into a plateau with further increasing strain. Such a linear stage is correlated to the elastic stress

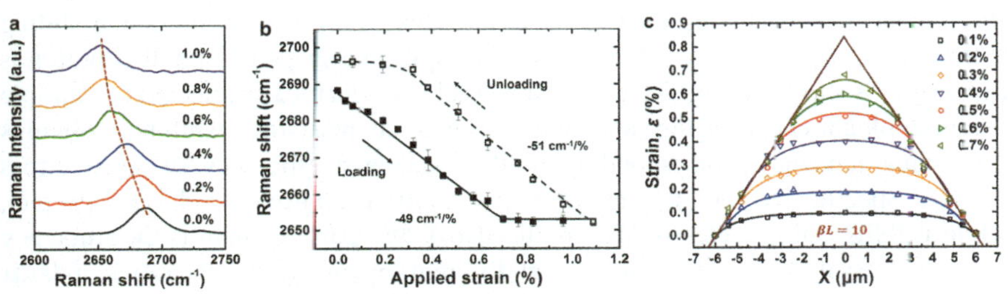

Fig. 2.4 **a** Evolution of the Raman 2D spectrum for graphene with increasing applied strain. **b** Raman 2D-band shifts during a loading/unloading cycle. **c** Strain distribution in the direction of stretching axis of graphene sheet at different strain levels

transfer from substrate to graphene sheet, followed by the occurrence of interfacial sliding evidenced by a plateau stage. Once the bending load is released, the peak position upshifts accordingly at a rate of -51 cm^{-1}/% and remain as a constant at the end of unloading. Conceivably, the graphene sheet tends to recover to its initial state with Raman 2D band located at ~2690 cm^{-1} as the strain is reduced to 0.4%. Further continuous release would cause slight compression as well as interfacial buckling supported by the relatively higher peak positions (~2695 cm^{-1}) at the second plateau.

To gain a deep understanding of the interfacial stress transfer from the PMMA substrate to graphene, and reveal the possible failure modes at the graphene/PMMA interface, a nonlinear shear-lag model was adopted as shown in Fig. 2.4c. In detail, it is clear that the strain builds up in graphene sheet from the edges along the x direction, and approaches a peak at the center, which is approximately equal to the strain in the matrix. At relatively low strain levels (e.g. $\varepsilon_m < 0.2\%$), the whole length of interface is in elastic conditions and the strain distribution could be fitted by Eq. (2.7) with $\beta L = 10$. Once sliding stress reaches the interfacial shear strength at higher strains, interfacial sliding occurs. Correspondingly, the sliding zone develops from both edges of the sheet, approaches the center gradually as the loading amplitude increases. Eventually, the in-plane strain distribution in graphene becomes almost linear and tends to be in a triangle-like shape near the edges. According to Eq. (2.9), the interfacial shear strength between graphene and PMMA was obtained as ~0.6 MPa, agreeing well with the results reported previously for PMMA/graphene/SU8 (0.3–0.8 MPa) and graphene/polyethylene terephthalate (PET) (0.7 MPa) laminates [20, 22].

2.4 Cyclic Loading of Graphene on Polymer Substrate

2.4.1 Interfacial Degradation

To examine the dynamic response of graphene/PMMA interface, cyclic loading/unloading tests were further conducted on the composite system through applying and releasing the bending load to the cantilever beam. Herein, the applied strain level is increased up to 0.7% to ensure the occurrence of interfacial sliding at the center of graphene sheet, and thus a triangle-like shaped strain distribution is expected. As shown in Fig. 2.5a, once the graphene sheet was subjected to multiple loading/unloading cycles, the slope of the strain gradient ($d\varepsilon/dx$) exhibits an obvious decreasing tendency, corresponding to the weakened efficiency of shear stress transfer. Specifically, for 1st loading process, $d\varepsilon/dx$ is approximated to 0.14%/μm, while at 20th loading cycles, $d\varepsilon/dx$ is reduced to 0.09%/μm. To quantify the efficiency of shear stress transfer across interface at different loading cycles, the ratios of $d\varepsilon/dx$ at a given cycle is normalized to that of at 1st cycle, $\eta = (d\varepsilon/dx)_{Nth}/(d\varepsilon/dx)_{1st}$. Figure 2.5b summarizes weakening trend of stress transfer efficiency during the twenty loading/unloading cycles, and a plateau is expected to further increase loading/unloading cycles. Such apparently decreased values imply the interface

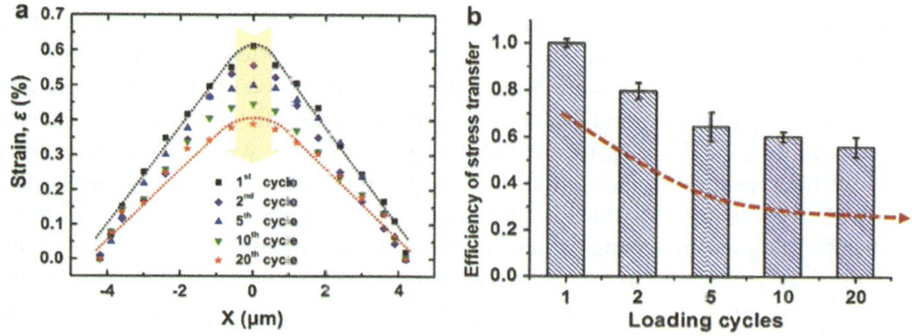

Fig. 2.5 **a** Strain distribution in the direction of stretching axis of graphene sheet at the strain level of 0.7% in multiple loading/unloading cycles. **b** Variation of the efficiency of stress transfer with increasing loading cycles

degradation once subjected to cyclic loads, which has not been considered in previous studies [45–47].

To gain more insights into the underlying degradation of interfacial shear stress transfer, AFM images were recorded to monitor the evolution of surface morphology of graphene sheet after each loading/unloading cycle. Herein, to quantitatively describe the surface morphology, the surface roughness as the most widely used roughness parameter was introduced, defined as $R_a = \frac{1}{n}\sum_{i=1}^{n}|y_i|$, where y_i is the vertical distance from the mean line to the ith data point in the AFM image. As shown in Fig. 2.6a, the graphene sheet is relatively flat with an average roughness of ~0.4 nm before the load is applied (calculated by the Nanoscope software). After the first loading/unloading cycle with tensile strain of 0.7%, the graphene flake presents randomly distributed buckles as shown in Fig. 2.6b. Correspondingly, the surface roughness of individual graphene sheet increases to ~1.73 nm. It is conceivable that the formation of buckles is related to the mechanical compression during unloading process as discussed above. It is worth noting that, different from the regular surface

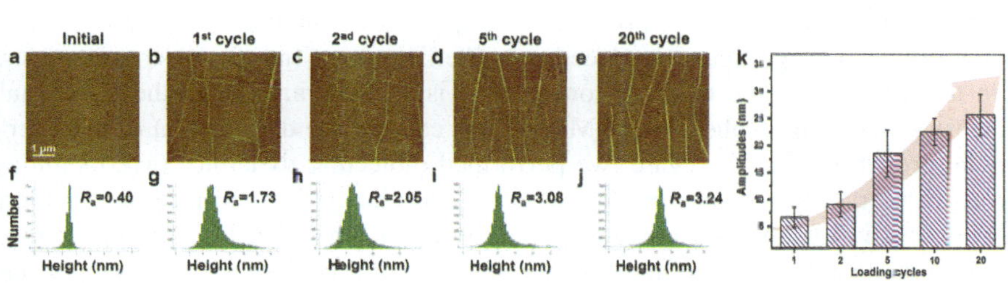

Fig. 2.6 **a–e** AFM images of the graphene sheet deposited on PMMA, which is subjected to twenty cycles of loading and unloading. The images are recorded from the same flake whereas the scanning area might slightly deviate from each other. The evolution of its morphology is demonstrated by the variation of roughness distribution as presented in **f–j**. **k** Increasing amplitudes of buckles with increasing loading cycles

morphology with periodic buckles in response to the uniaxial compression in most systems [48, 49], the formation of buckles with random orientation here could be attributed to the Poisson ratio effect during the loading/unloading cycles, as well as the influence of geometry and orientation of individual graphene sheets with respect to the loading axis [50, 51]. Compared with the 1st cycle, we find that the amplitudes of buckles tend to increase in the 2nd loading/unloading cycle as presented in Fig. 2.6k. Accordingly, the surface roughness also increases to ~2.05 nm. With further bending and releasing cycles up to the 5th, the amplitudes of buckles grow continuously together with the enhancement of averaged surface roughness ~3.08 nm. After 20th loading/unloading cycles, merely slight increase in amplitudes could be observed and the corresponding surface roughness is improved to ~3.24 nm. This sluggishly increased tendency is in good agreement with the plateau stage following the degradation of interfacial shear stress presented in Fig. 2.6b.

We suggest that the formation of buckles after mechanical loading/unloading might break the full contact between graphene and PMMA into patches. On one hand, the decreased contact area would impair the stress transfer at the graphene/PMMA interface. As we know, the stress build-up in fillers requires a transition region corresponding to the critical length (L_c) typically defined as twice the distance over which the strain rises to 90% of the plateau level, which determines the reinforcing effect of composite material [20, 45]. As shown in Fig. 2.7a, the loss of the stress transfer hence happens in the transition region (S_1) while S_2 region is responsible for the effective load bearing. The stress transfer efficiency is thus determined by

$$\eta = \frac{S_2}{2S_1 + S_2} = \frac{\int_0^L \varepsilon(x)\mathrm{d}x}{\varepsilon_m L} = 1 - \frac{\tanh(\beta L/2)}{\beta L/2} \tag{2.12}$$

Take an example from Fig. 2.4c, the length of graphene flake is $L = 15$ μm, and the fitting parameter is $\beta L = 10$, yielding $\eta = 0.8$. After the formation of buckles, the contact between graphene and PMMA splits into N patches, which is assumed to be in equal size for simplification. AFM imaging shows the length of each patch of contact as $L_a = 1.5$ μm and the lateral span of each buckle as $L_b = 0.1$ μm. Consequently, the stress transfer efficiency is approximated as $\eta = N[1 - \tanh(\beta L_a/2)/(\beta L_a/2)] = 0.55$, suggesting significantly reduced overall efficiency of interfacial stress transfer. On the other hand, the reduced contact area is also detrimental to the interfacial adhesion between graphene and PMMA, thus causing the degradation of the interfacial properties. In this regard, we performed molecular dynamics (MD) simulations to extract key mechanical parameters and clarify the interface degradation during cyclic loading and unloading. As shown in Fig. 2.7c, initially, the monolayer graphene appears flat and keeps a relatively full contact with the supporting PMMA substrate. Then, the graphene sheet is strained through shear stress across graphene/PMMA interface. The interfacial sliding occurs as the matrix strain is higher than 3%, and then the average strain in graphene would step into a plateau region (Fig. 2.7d-ii). Meanwhile, the shear stress reaches τ_c and the interfacial stress transfer is governed by interfacial sliding. Upon unloading, the strain in graphene

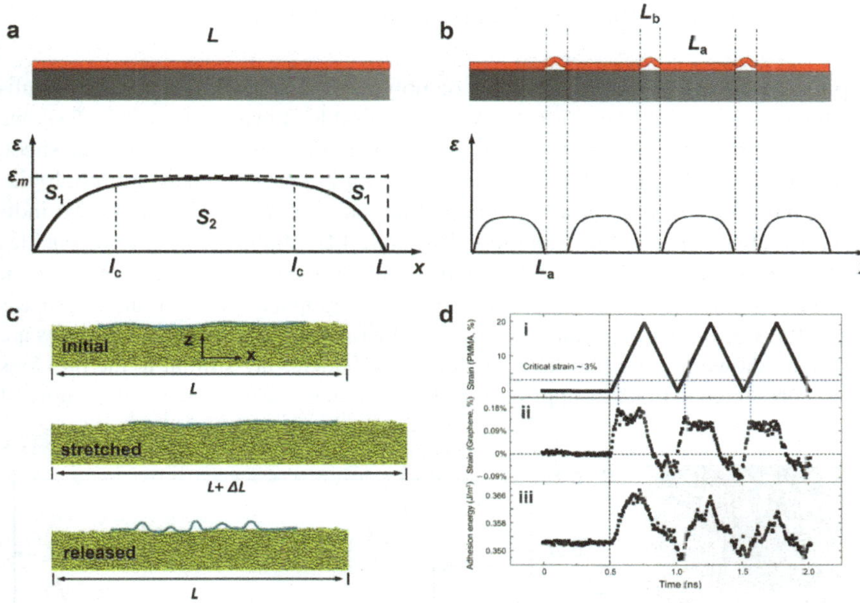

Fig. 2.7 Schematic plots of strain distribution within the graphene sheet with **a** full and **b** partial contact with the substrate, respectively. **c** Simulation snapshots that illustrate the morphological changes of graphene from a flat sheet fully in contact with the substrate to a series of partial contact separated by buckles formed due to the compression-induced buckling upon unloading. **d** The evolution of strain in the substrate, graphene and their interfacial adhesion energy as a function of time

drops immediately, even down to the negative value, implying the subtle compressive strain it bears. Consequently, the graphene sheet tends to buckle to release the compressive strain energy. The weakened interfacial adhesion and efficiency of shear stress transfer are expected due to the reduced interfacial contact area during the subsequent loading cycles. Figure 2.7d-iii summarizes the variation of average strain in graphene and the interfacial adhesion energy over three cycles. In the 2nd and 3rd loading/unloading cycles, both the average strain in graphene and the interfacial adhesion energy degrade apparently due to interfacial sliding and the subsequent buckles induced by compression. Note that, due to the discrepancy of graphene sheet sizes in simulation and experimental setups (~40 nm vs. ~10 μm), the absolute strain level in graphene would be different. Even the limited spatial resolution of Raman spectroscopy (~1 μm) hinders us from directly visualizing such variation in terms of strain distribution of the stretched graphene, the theoretical analysis together with MD simulations offers straightforward evidence upon the combined effect of reduced contact area and decreased interfacial adhesion which is responsible for the observed degradation of stress transfer at the graphene/PMMA interface.

2.4.2 Thermally-Induced Mechanical Recovery

Inspired by the shape memory effect of polymeric materials that is the stimuli-responsive control of transition between metastable morphologies [52, 53], we employ thermal treatment to modulate the buckled morphologies of graphene sheet. AFM images were recorded to characterize the variation of graphene morphologies before and after mechanical and thermal treatment. Figure 2.8a presents two individual graphene sheets with ribbon and triangle-like shapes with several buckles parallelized to each other within the relatively flat graphene sheets. The presence of buckles was attributed to the residual strain during graphene sheets transfer process [43]. After twenty loading/unloading cycles, the randomly oriented new buckles are formed as shown in Fig. 2.8b. The amplitudes of buckles are similar to the previous ones as presented in Fig. 2.6, approximated to 6.8 nm, while much lower than those of

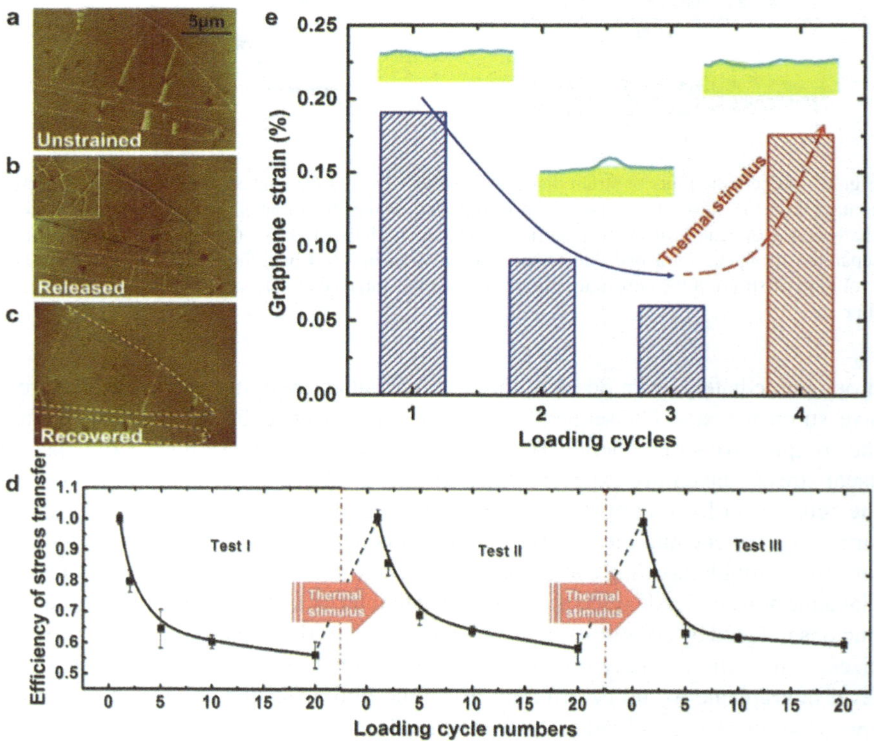

Fig. 2.8 AFM images of graphene **a** before and **b** after cycling loads, and **c** after thermal treatment. **d** The recovery of degraded interfacial stress transfer triggered by thermal stimuli. **e** The recovery of the average strain in graphene by the thermal stimulus. Insets are simulation snapshots of a graphene sheet deposited on to the PMMA substrate in the unstrained, released and recovered states, respectively

pre-exist buckles (~23.2 nm). Once the thermal treatment was applied to the sample (at ~40 °C for a week), the enhanced mobility of polymer chains under thermal stimuli would facilitate the flattening of graphene and hence relaxation of buckles (Fig. 2.8c). Meanwhile, we notice that the thermal stress applied to the graphene sheet could be neglected, where negligible difference (3 cm^{-1}) in the Raman 2D band frequency before and after thermal treatment was observed.

To experimentally validate the recovery of the interfacial interaction, cyclic mechanical tests were applied to the same specimen and the efficiency of stress transfer was calculated for each degradation/recovery cycle. Herein, we name the first degradation/recovery cycle as a Test I as plotted in Fig. 2.8d, which suffered 20 loading/unloading cycles and followed by the thermal treatment. Obviously, the efficiency of shear stress transfer extracted from Test II almost recovers to the initial value, and then repeats the similar decreasing tendency in the subsequent loading/unloading cycles. Besides, similar shear stress transfer behaviors are observed in the Test III.

To explore the reconstruction of graphene/PMMA interfacial interaction under thermal stimuli in molecular simulation, a three-cycle loading condition was simulated at ultra-low temperature (0 K) as the system in simulation is more sensitive to thermal fluctuation than that in experiments. As a result, obvious buckle configurations and relatively larger reduction of strain in graphene are observed in Fig. 2.8e. After three-cycle loading, thermal stimulus of 300 K heating up is introduced to the system, and the strain in graphene can be remarkably recovered through thermally activated interfacial sliding that heal the buckles, with a reconstructed interfacial morphology and interaction. Interestingly, the recovered interfacial performance in the graphene/PMMA composites is apparently different from that of in conventional fiber reinforced composites. This highlights the uniqueness of graphene-based nanoscale interface, where the weak vdW interactions break and reform during the shear deformation to keep the efficient load transfer, thus granting the interface with the impressing mechanical recoverability [28, 29, 54, 55].

2.5 Interfacial Engineering via Chemical Functionalization

2.5.1 In Situ Oxidation of Graphene

In light of the vdW nature of graphene/PMMA interface, we carried out the ozone treatment to graphene and introduced oxygen-containing functional groups on its surface [56], where the pendant hydroxyl groups tend to form hydrogen bonding with carbonyl groups of PMMA (Fig. 2.9a) [19, 57, 58]. Figure 2.9b shows the typical FTIR spectra of the graphene sheets treated by O_3/H_2O gaseous mixture at different times. As expected, no apparent FTIR bands could be observed for pristine graphene sheet due to its infrared inactive carbon–carbon double bonds [59]. After oxidation treatment, the peak at 1720 cm^{-1} related to carboxylic groups, a broad

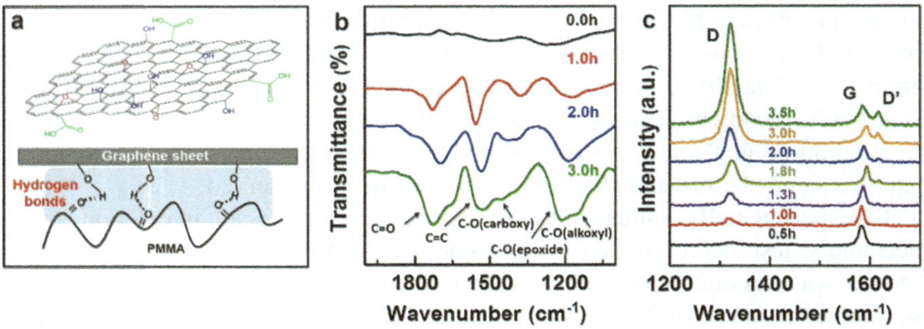

Fig. 2.9 **a** Schematic of an oxidized graphene sheet having various oxygen-containing groups and interacting with the carbonyl groups of PMMA through hydrogen bonds. **b** FTIR spectra and **c** Raman spectra of oxidized graphene treated at different reaction times

band around 1200 cm^{-1} assigned to the coupling of ν(C–O) of alkoxyl and epoxide groups, and the stretching mode of ν(C=C) at 1580 cm^{-1} in the graphene backbone are observed [19]. With further increasing oxidization time up to 3 h, the band intensity of carboxylic groups at 1720 cm^{-1} relative to that of C=C backbone at 1580 cm^{-1} tended to increase, indicating an enhancement of oxygen-containing groups in the oxidized graphene samples. Meanwhile, the apparent Raman D-band and D'-band are observed for the oxidized graphene samples, implying the conversion of sp^2 carbon to sp^3 carbon due to the oxygen-containing functional groups grafted onto graphene sheet [60]. Figure 2.9c shows the evolution of Raman D-band and D'-band intensity in the consecutive spectra, in which the continuous attachment of oxygenated groups onto graphene sheet leads to significant changes in the relative intensity of the D-band normalized to G band. In the following analysis, we used Raman intensity ratio of I_D/I_G to identify the functionalized degrees of oxidized graphene sheets at a certain range [61].

2.5.2 *Interfacial Properties Dominated by Hydrogen Bonding*

The formation of the hydrogen bonding (H-bonding) is anticipated to enhance the interfacial shear strength in graphene/PMMA nanocomposite system. In order to investigate the interfacial property between oxidized monolayer graphene sheet and PMMA substrate, the in situ Raman measurement was performed with the cantilever beam loading. It is noted that, the H-bonding features breaking-and-reforming characteristics so that interfacial slippage failure can also be expected. Figure 2.10a shows the strain distribution of oxidized monolayer graphene sheet along the dashed line at different applied strain levels. It can be seen from the fitting results that both the fitted βL values and the corresponding slopes $d\varepsilon/dx$ for oxidized graphene sheet are higher than those of pristine graphene sheet with similar lateral sizes ($\beta = 2.5$, $d\varepsilon/dx = 0.35\%/\mu m$ vs. $\beta = 0.6$, $d\varepsilon/dx = 0.16\%/\mu m$). This is a clear indication of the better

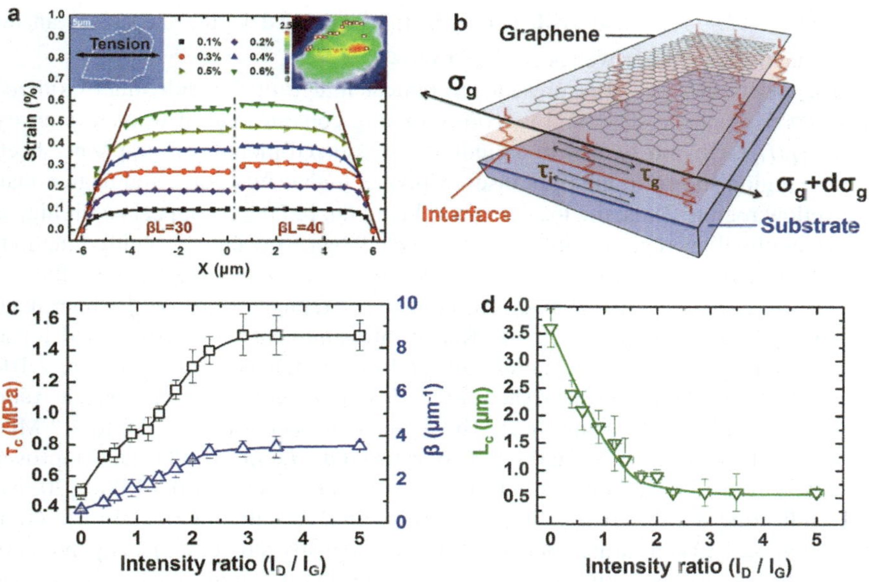

Fig. 2.10 **a** Strain distribution in the direction of the tensile axis of the oxidized graphene sheet at different strain levels. The insets are (left panel) optical image of tested graphene sheets and (right panel) its corresponding Raman contour map of I_D/I_G. **b** Schematic of the interface layer with the interactions modeled by linear springs. **c** The interfacial shear strength (τ_c) and stress transfer factor (β) as well as **d** the critical length (l_c) as a function of the Raman intensity ratio of I_D/I_G in graphene/PMMA system

stress transfer performance obtained for oxidized monolayer graphene sheet, which could be attributed to the H-bonding at interface.

The interfacial strengthening can also be well explained by a modified shear-lag model. In the case of functionalized interface, the stress transfer efficiency is not only determined by the stiffness of matrix and graphene, but also the stiffness of the interface. As shown in Fig. 2.10b, we introduced the effect of interface layer in the shear-lag analysis. The displacement of the interface is u_i and satisfies

$$\tau_g = k_i\left(u_i - u_g\right) \tag{2.13}$$

$$\tau_i = k_m(u_m - u_i) \tag{2.14}$$

τ_g is the shear stress between graphene and interface layer, τ_i is the shear stress between polymer matrix and interface layer, corresponding to the stiffness of k_i and k_m respectively. The force balance at the interface layer requires

$$\tau_g = \tau_i = k\left(u_m - u_g\right) \tag{2.15}$$

where $k = \frac{k_i k_m}{k_i + k_m}$. As k_i rises with increasing functionalized degree, the stress transfer efficiency is hence improved according to $\beta = \sqrt{k/E_g t}$.

The insets in Fig. 2.10a present the optical image of the individual oxidized monolayer graphene sheet and its corresponding contour map of Raman intensity ratio (I_D/I_G). Obviously, within an individual oxidized monolayer graphene sheet, the red region shown in the right inset exhibits a higher functionalized degree than the yellow region. Thus, the local strain value in the red region appears a bit higher than that of yellow region, implying a relatively larger βL value for highly functionalized part ($\beta L = 40$ vs. $\beta L = 30$). In order to systematically clarify the influence of functionalized degrees of oxidized monolayer graphene sheet on the interfacial property of nanocomposites, in situ Raman measurements were performed for at least 9 samples with various lateral sizes and functionalized degrees. Figure 2.10c shows the trend of the interfacial shear strength as a function of Raman intensity ratio (I_D/I_G) of oxidized graphene sheet. The τ_c tends to increase from 0.6 to 1.7 MPa linearly with increasing Raman intensity ratio of I_D/I_G up to ~3.0 and improves by nearly four times compared with previous results (averaged 0.4 MPa) [20, 41, 43], indicating that the H-bonding at interface enhances the interfacial interaction unambiguously. With further increasing Raman intensity ratio of I_D/I_G beyond ~3.0, while, the τ_c reaches a plateau region, indicating that the interfacial interaction is hard to improve any more. Similar phenomenon is also observed for critical length (L_c) which shows a notable decline from 5.4 to 1.2 μm, followed by a plateau stage as presented in Fig. 2.10d. Clearly, the reduced L_c is corresponding to higher βL and hence the enhanced stress transfer efficiency, which is resulted from the stronger interfacial interaction.

As mentioned earlier that the interfacial interaction between pristine graphene sheet and underlying polymer substrate is only vdW force and it is not strong enough to ensure efficient stress transfer. Based on our work, in efforts to more effectively take advantage of the high Young's modulus and strength of graphene, it is essential to attach oxygenated groups onto graphene sheet to strengthen the interfacial interaction between graphene and PMMA substrate via H-bonding. However, it is also noteworthy that the interfacial shear strength gets saturated with further increasing numbers of oxygenated groups, implying the optimized functionalized degree is required to maximize the interfacial adhesion. Zhang et al. endowed graphene with hydrogen and oxygen functionalization of 3% in MD simulations, and observed about 48 and 183% increase of interfacial shear force respectively, compared to pristine monolayer graphene. Further increase of oxygen coverage up to about 7% led to a saturated interfacial shear force, which is aligned with our experimental results [62]. Otherwise, the excess oxygenated groups grafted onto graphene sheet might create defects in the graphene lattice, and then compromise the mechanical properties of individual graphene sheet itself. A critical issue hence arises regarding the optimization of functionalized degree to increase the interfacial shear strength without compromising intrinsic mechanical properties of graphene sheet in nanocomposites.

2.5.3 Failure Modes at Functionalized Interface

Since frictional behaver occurs at graphene/PMMA interfaces, matrix will create the compressive restraints on graphene sheets after releasing the applied strain, leading to wrinkling or buckle delamination in order to release the compressive strain [63]. In contrast to pristine graphene, the distribution of buckle ridges tends to be denser accompanied with the presence of periodic wrinkles, inducing a highly non-uniform strain fields in graphene sheets. This is evidenced by the Raman line-mapping that shows an undulating variation of local strains at the center of graphene flake (Fig. 2.11a). The appearance of wrinkles also indicates of a stronger interfacial adhesion [64]. Based on the graphene buckle height (h) and width (w) measured by AFM, adhesion energy of graphene-based interfaces can be calculated by a strain-independent formula [65]:

$$\Gamma = \frac{E_g t}{2}\left(\frac{\pi h}{2w}\right)^4 + 8D\left(\frac{\pi^2 h}{2w^2}\right)^2 \tag{2.16}$$

where $E_g = 1$ TPa is the Young's modulus of graphene, $t = 0.335$ nm is the graphene thickness, $D = \frac{E_g t^3}{12(1-\upsilon^2)}$ is the bending rigidity of graphene and $\upsilon = 0.16$ is the Poisson's ratio. The buckle height and width for pristine graphene-based are $h =$ 5.3 nm and $w = 145$ nm respectively, which increases to $h = 15.2$ nm, $w = 189$ nm for functionalized interface. Accordingly, the adhesion energy is also enhanced from 0.4 to 32.2 mJ/m^2, two orders of magnitude higher due to the stronger H-bonding.

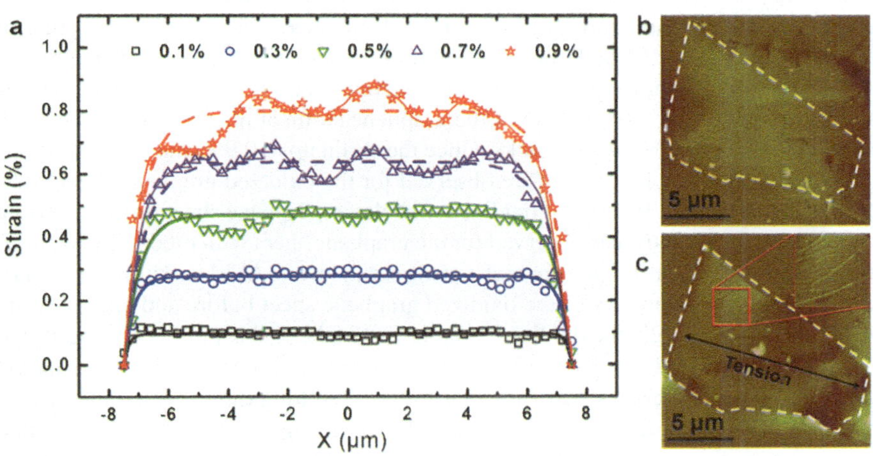

Fig. 2.11 **a** Strain distribution of the oxidized graphene sheet (I_D/I_G ~ 3.5) along the stretching direction at different strain levels. AFM images present the relatively flat (**b**) and buckled morphology (**c**) of graphene sheet before the tension and after the release

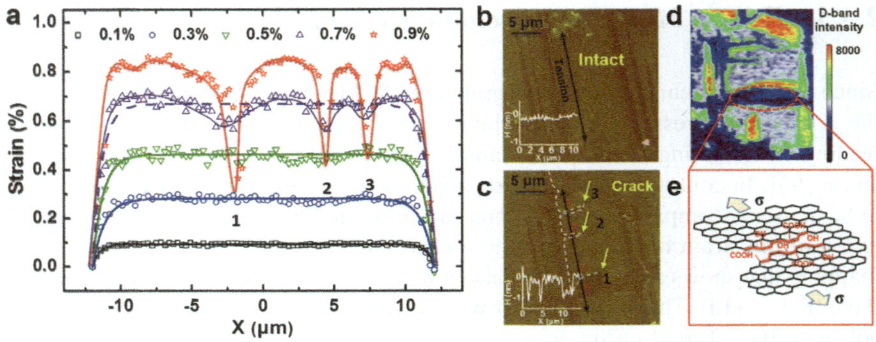

Fig. 2.12 **a** Strain distribution of monolayer graphene with I_D/I_G ~ 5.0 at different strain levels. Cracks tend to occur at high strain levels. AFM images of oxidized graphene sheet **b** before the tension and **c** after the release; the insets show the height profile of the tested graphene sheet. **d** Raman image of D-band intensity of oxidized graphene sheet after tension. **e** Schematic illustrating the crack growth behavior of graphene provoked from tensile stress

Note that, the excessive oxygen-containing groups would create defects in the graphene lattice and induce the crack initiation during the tension. Meanwhile, it may also weaken the mechanical properties of individual graphene sheet [66]. To experimentally clarify such influence of excessively functionalized graphene sheet on the interfacial properties as well as failure mode of individual graphene sheet, herein, we monitor the strain distribution of highly oxidized individual graphene sheet with I_D/I_G ~ 5.0 at various strain levels. At lower strain level, the interfacial shear stress can be efficiently transferred from matrix to the individual graphene sheet, and the plateau strain of oxidized graphene sheet carried is close to that of matrix shown in Fig. 2.12a. However, with further increasing strain level up to 0.7%, unlike the sliding behavior at the edges of moderate functionalized graphene sheet as shown in Fig. 2.10a, a slight drop of the graphene strain at plateau region occurred, indicating the initiation of the cracks. Once the strain up to 0.9%, the apparent drops of graphene strain to low values are observed for the oxidized graphene sheet, indicating the permanent fracture of oxidized monolayer graphene sheet. In comparison, no apparent drop of strain is observed for the graphene sheet with moderate functionalized degrees even suffered from high tensile strain levels. AFM was utilized to characterize the damage modes of the oxidized graphene sheet before and after tension. As shown in Fig. 2.12b, the oxidized monolayer graphene sheet initially presents the intactness with relatively flat surface. Once subjected to axial tensile deformation up to 0.9%, several cracks perpendicular to the stretching direction are inclined to propagate in the oxidized graphene sheet shown in Fig. 2.12c. The height profile across the cracks pronouncedly displays the gaps at three marked sites, which was consistent with the obvious graphene strain drop sites as shown in Fig. 2.12a. Additionally, the Raman image of D-band intensity in Fig. 2.12d after tensile deformation reveals that the cracks are prone to form at the sites with higher D-band intensity, implying that the excess oxygen-containing groups in the oxidized graphene sheet

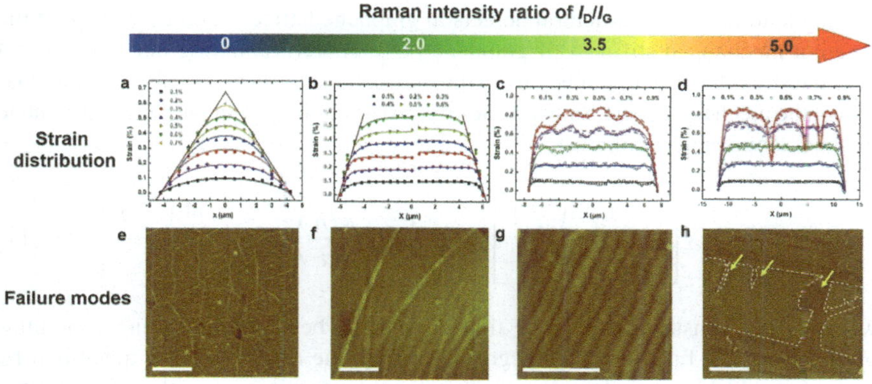

Fig. 2.13 **a–d** Evolution of strain distribution of graphene sheets along the tensile direction with increasing Raman I_D/I_G ratio, and **e–h** corresponding failure modes of graphene sheets after unloading as characterized by AFM: buckling, coexistence of buckling and wrinkling, wrinkling, fracture. Scale bars in **e–g** are 1 μm and scale bar in **h** is 5 μm

acted as defect sites to induce the initiation of cracks. The fracture mechanism of oxidized graphene sheet could be further elucidated by the schematic drawing as shown in Fig. 2.12e. The various oxygen-containing groups such as OH, C–O–C, or COOH grafted onto graphene sheet during oxidation process would lead to naturally configuration changes caused by orbital hybridization from a planar sp^2-hybridized geometry to a distorted sp^3-hybridized geometry [67, 68]. Once external load is applied to the highly oxygenated graphene sheet, the initiated or pre-existing defects are propagated perpendicular to the loading direction, and resulted in catastrophic failure of the graphene sheet.

Depending on the functionalized degrees of oxidized graphene sheet, distinct failure modes including the interfacial shear sliding, buckling, wrinkling and fracture of graphene sheet are identified and summarized in Fig. 2.13. Such a systematic investigation in the evolution of deformation modes of graphene fillers as a function of the functionalized degrees offers valuable insight and design guideline for implementing reinforcing effects in nanocomposites.

2.5.4 Implications for Graphene-Reinforced Nanocomposite

From the perspective of reinforcing mechanism, the nanofiller size and interfacial strength are two dominant factors that determine the load-bearing capacity of graphene. Generally, the larger the lateral dimension of graphene flake and the higher the interfacial strength, the more effective load transfer to enhance the mechanical properties of graphene-based nanocomposite [45]. As discussed above, chemical functionalization is an effective way to improve the interface strength, while it is inevitable to introduce structural defects in graphene lattices, thereby damaging the inherent mechanical stiffness of graphene [69]. Therefore, taking into account not only the stress transfer efficiency from the matrix to nanofillers, but also the degradation of the mechanical performance, the effective modulus of functionalized graphene in nanocomposite is defined as:

$$E_{eff} = E_{defect}\left[1 - \frac{\tanh\left(\frac{\beta L}{2}\right)}{\frac{\beta L}{2}}\right] = E_g\left(1 - \frac{kI_D}{I_G}\right)\left[1 - \frac{\tanh\left(\frac{\beta L}{2}\right)}{\frac{\beta L}{2}}\right] \tag{2.17}$$

where k is a constant to describe the relationship between the Young's modulus of graphene and functionalized degree. Actually, the k value can be quantified by performing AFM nanoindentation tests on monolayer graphene with different functionalized degrees. Here, we assume a linear reduction of graphene modulus with increasing Raman I_D/I_G ratio and take $k = 0.1$ for simplification, which will not affect the conclusions we draw.

As shown in Fig. 2.14a, increasing the flake dimensions and strengthening the interface can both improve the effective modulus of graphene. In particular, the effect of flake dimension is explained in detail by the shear lag, where a critical length exists for effective reinforcement. The orange curve ($\beta = 0.5$) also validates the statement that the minimum critical length for reinforcement in a model graphene/PMMA system is on the order of ~3 μm. In contrast, it is generally thought that an optimal length of 30 μm would be required for good reinforcement, allowing the stress in graphene almost identical to that applied in matrix [20, 45]. As an example, Liu et al. employed a layer-by-layer method to construct a large composite stack with CVD graphene spanning the physical dimensions of a surrounding matrix, approaching a limit of maximal mechanical reinforcement at minimal platelet addition per mass of material [70]. On the other hand, it is found that strengthening of interface tends to be saturated with increasing functionalized degree (corresponding to increased β),

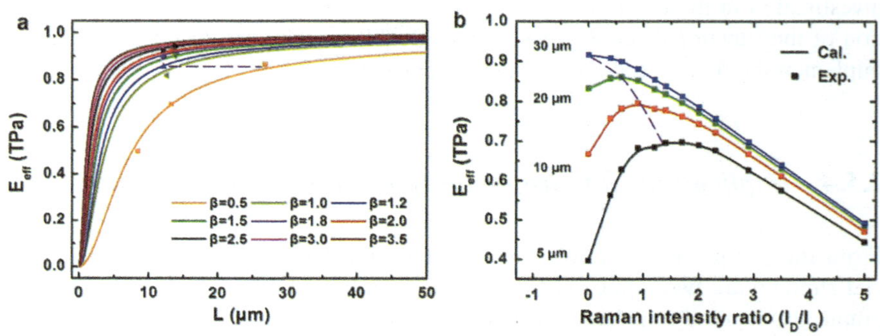

Fig. 2.14 The variation of effective modulus of oxidized graphene in the nanocomposite as a function of **a** the flake length with different stress transfer factors and **b** Raman intensity ratio (I_D/I_G) with different given lengths of graphene sheets

while the growing defects in graphene will cause the degradation in the effective modulus as presented in Fig. 2.14b.

In fact, in addition to the flake dimension and interface strength, other factors including the dispersion, orientation and waviness, also contribute to the reinforcing mechanism in nanocomposites [71]. Recently, Papageorgiou and Young et al. have reviewed all these parameters and introduced corresponding coefficients into the effective modulus of graphene (E_{eff}) such as the Krenchel orientation factor that is taken as $\eta_O = 8/15$ for randomly oriented nanofiller [72, 73]. Besides, the fabrication of layered graphene nanocomposite with ultrahigh orderliness and alignment has demonstrated a highly efficient load transfer between the polymer phase and the stiff graphene nanoplatelets, leading to exceptional mechanical properties of nanocomposite films [74, 75].

It should be noted that, herein, graphene is deposited on the surface of the polymer substrate, instead of being embedded in matrix. Apparently, physical constraints will be imposed on embedded graphene from both sides within the matrix, which will definitely affect the mechanical response of the interface during deformation. For instance, the buckling instability formed during the cyclic loading process could be greatly suppressed. However, the microscale mechanisms of stress transfer and interfacial failure are the same for two cases. In fact, previous studies demonstrated that the Raman 2D-band shift rate for the graphene monolayer is similar with and without the polymer top coat [42]. Therefore, the exposure of graphene enables in situ AFM imaging to better monitor the interfacial deformation characteristics and reveal failure mechanism at the nanoscale.

2.6 Chapter Summary

We demonstrate that a combination of in situ tensile micro-Raman technique and AFM measurements offers a thorough insight into the mechanical behavior at graphene/PMMA interfaces. Dominated by the weak vdW interactions, the interfacial shear strength is only on the order of 0.6 MPa, which easily induces the interfacial slip of graphene during loading. Upon unloading, the compression leads to the formation of buckling that breaks the full contact between graphene and PMMA into patches. The impaired contact reduces the adhesion and stress transfer efficiency at interface, accounting for the further degradation under cyclic loading. In addition to the mechanical recovery via thermal treatment, interfacial engineering through chemical functionalization is also effective to resolve the degradation issues. The introduction of strong hydrogen bonding significantly strengthens the interface, with the interfacial shear strength approximately tripled that of pristine graphene. Furthermore, the toughness of composite materials will also benefit from its reformable nature. However, as the excessive functionalization creates defects and cracking in graphene sheets, there exists therefore a balance to be struck in the design of graphene-based nanocomposites between the ability to achieve higher interfacial adhesion and the reduction in the elastic modulus of the functionalized graphene sheet.

References

1. Lee C, Wei X, Kysar JW, Hone J (2008) Measurement of the elastic properties and intrinsic strength of monolayer graphene. Science 321(5887):385–388
2. Wei Y, Yang R (2019) Nanomechanics of graphene. Natl Sci Rev 6(2):324–348
3. Zhiping X, Zheng Q (2018) Micro- and nano-mechanics in China: a brief review of recent progress and perspectives. Sci China Phys Mech Astron 61(7):074601–074614
4. Cao K, Feng S, Han Y, Gao L, Hue Ly T, Xu Z et al (2020) Elastic straining of free-standing monolayer graphene. Nat Commun 11(1):284
5. Liu Q, Liu L, Xie K, Meng Y, Wu H, Wang G et al (2015) Synergistic effect of a r-GO/PANI nanocomposite electrode based air working ionic actuator with a large actuation stroke and long-term durability. J Mater Chem A 3(16):8380–8388
6. Owais M, Zhao J, Imani A, Wang G, Zhang H, Zhang Z (2019) Synergetic effect of hybrid fillers of boron nitride, graphene nanoplatelets, and short carbon fibers for enhanced thermal conductivity and electrical resistivity of epoxy nanocomposites. Compos Part A Appl Sci Manuf 117:11–22
7. Zhang Y, Liu L, Sun B, Wang G, Zhang Z (2016) Preparation of lipophilic graphene oxide derivates via a concise route and its mechanical reinforcement in thermoplastic polyurethane. Compos Sci Technol 134:36–42
8. Dai Z, Wang G, Liu L, Hou Y, Wei Y, Zhang Z (2016) Mechanical behavior and properties of hydrogen bonded graphene/polymer nano-interfaces. Compos Sci Technol 136:1–9
9. Wagner HD, Vaia RA (2004) Nanocomposites: issues at the interface. Mater Today 7(11):38–42
10. Barber AH, Cohen SR, Eitan A, Schadler LS, Wagner HD (2006) Fracture transitions at a carbon-nanotube/polymer interface. Adv Mater 18(1):83–87
11. Evans A, Zok F, McMeeking R (1995) Fatigue of ceramic matrix composites. Acta Metall Mater 43(3):859–875
12. Cho CD, Holmes JW, Barber JR (1991) Estimation of interfacial shear in ceramic composites from frictional heating measurements. J Am Ceram Soc 74(11):2802–2808
13. Pupurs A, Goutianos S, Brondsted P, Varna J (2013) Interface debond crack growth in tension-tension cyclic loading of single fiber polymer composites. Compos Part A Appl Sci Manuf 44:86–94
14. Tang L-C, Zhang H, Wu X-P, Zhang Z (2011) A novel failure analysis of multi-walled carbon nanotubes in epoxy matrix. Polymer 52(9):2070–2074
15. Chou CT, Gaur U, Miller B (1993) Fracture mechanisms during fiber pull-out for carbon-fiber-reinforced thermosetting composites. Compos Sci Technol 48(1–4):307–316
16. Gao Y, Li JZ, Liu LQ, Ma WJ, Zhou WY, Xie SS et al (2010) Axial compression of hierarchically structured carbon nanotube fiber embedded in epoxy. Adv Funct Mater 20(21):3797–3803
17. Ma W, Liu L, Yang R, Zhang T, Zhang Z, Song L et al (2009) Monitoring a micromechanical process in macroscale carbon nanotube films and fibers. Adv Mater 21(5):603–608
18. Ma W, Liu L, Zhang Z, Yang R, Liu G, Zhang T et al (2009) High-strength composite fibers: realizing true potential of carbon nanotubes in polymer matrix through continuous reticulate architecture and molecular level couplings. Nano Lett 9(8):2855–2861
19. Gao Y, Liu L-Q, Zu S-Z, Peng K, Zhou D, Han B-H et al (2011) The effect of interlayer adhesion on the mechanical behaviors of macroscopic graphene oxide papers. ACS Nano 5(3):2134–2141
20. Gong L, Kinloch IA, Young RJ, Riaz I, Jalil R, Novoselov KS (2010) Interfacial stress transfer in a graphene monolayer nanocomposite. Adv Mater 22(24):2694–2697
21. Pugno NM, Yin Q, Shi X, Capozza R (2013) A generalization of the Coulomb's friction law: from graphene to macroscale. Meccanica 48(8):1845–1851
22. Jiang T, Huang R, Zhu Y (2014) Interfacial sliding and buckling of monolayer graphene on a stretchable substrate. Adv Funct Mater 24(3):396–402
23. Dai ZH, Wang YL, Liu LQ, Liu XL, Tan PH, Xu ZP et al (2016) Hierarchical graphene-based films with dynamic self-stiffening for biomimetic artificial muscle. Adv Funct Mater 26(38):7003–7010

24. Chen Y, Dai Z, Weng C, Wang G, Liu X, Cong X et al (2018) Engineering the interface in mechanically responsive graphene-based films. RSC Adv 8(63):36257–36263
25. Wan S, Chen Y, Wang Y, Li G, Wang G, Liu L et al (2019) Ultrastrong graphene films via long-chain π-bridging. Matter 1(2):389–401
26. Wan S, Li Y, Mu J, Aliev AE, Fang S, Kotov NA et al (2018) Sequentially bridged graphene sheets with high strength, toughness, and electrical conductivity. Proc Natl Acad Sci USA 115(21):5359–5364
27. Cheng Q, Duan J, Zhang Q, Jiang L (2015) Learning from nature: constructing integrated graphene-based artificial nacre. ACS Nano 9(3):2231–2234
28. Liu Y, Xu Z (2014) Multimodal and self-healable interfaces enable strong and tough graphene-derived materials. J Mech Phys Solids 70:30–41
29. Zhu H, Zhu S, Jia Z, Parvinian S, Li Y, Vaaland O et al (2015) Anomalous scaling law of strength and toughness of cellulose nanopaper. Proc Natl Acad Sci 112(29):8971–8976
30. Liu Y, Xie B, Zhang Z, Zheng Q, Xu Z (2012) Mechanical properties of graphene papers. J Mech Phys Solids 60(4):591–605
31. Medhekar NV, Ramasubramaniam A, Ruoff RS, Shenoy VB (2010) Hydrogen bond networks in graphene oxide composite paper: structure and mechanical properties. ACS Nano 4(4):2300–2306
32. Compton OC, Cranford SW, Putz KW, An Z, Brinson LC, Buehler MJ, Nguyen ST (2012) Tuning the mechanical properties of graphene oxide paper and its associated polymer nanocomposites by controlling cooperative intersheet hydrogen bonding. ACS Nano 6(3):2008–2019
33. Najafi F, Wang G, Mukherjee S, Cui T, Filleter T, Singh CV (2020) Toughening of graphene-based polymer nanocomposites via tuning chemical functionalization. Compos Sci Technol 194:108140
34. Wang G, Dai Z, Liu L, Hu H, Dai Q, Zhang Z (2016) Tuning the interfacial mechanical behaviors of monolayer graphene/PMMA nanocomposites. ACS Appl Mater Interfaces 8(34):22554–22562
35. Wang G, Gao E, Dai Z, Liu L, Xu Z, Zhang Z (2017) Degradation and recovery of graphene/polymer interfaces under cyclic mechanical loading. Compos Sci Technol 149(8):220–227
36. Peng K, Liu L-Q, Li H, Meyer H, Zhang Z (2011) Room temperature functionalization of carbon nanotubes using an ozone/water vapor mixture. Carbon 49(1):70–76
37. Timoshenko S, Gere JM (2012) Theory of elastic stability. Dover Publications
38. Dobb MG, Johnson DJ, Park CR (1990) Compressional behavior of carbon-fibers. J Mater Sci 25(2):829–834
39. Androulidakis C, Koukaras EN, Frank O, Tsoukleri G, Sfyris D, Parthenios J et al (2014) Failure processes in embedded monolayer graphene under axial compression. Sci Rep 4:5271
40. Cox H (1952) The elasticity and strength of paper and other fibrous materials. Br J Appl Phys 3(3):72
41. Young RJ, Gong L, Kinloch IA, Riaz I, Jalil R, Novoselov KS (2011) Strain mapping in a graphene monolayer nanocomposite. ACS Nano 5(4):3079–3084
42. Gong L, Young RJ, Kinloch IA, Riaz I, Jalil R, Novoselov KS (2012) Optimizing the reinforcement of polymer-based nanocomposites by graphene. ACS Nano 6(3):2086–2095
43. Anagnostopoulos G, Androulidakis C, Koukaras EN, Tsoukleri G, Polyzos I, Parthenios J et al (2015) Stress transfer mechanisms at the submicron level for graphene/polymer systems. ACS Appl Mater Interfaces 7(7):4216–4223
44. Tsoukleri G, Parthenios J, Papagelis K, Jalil R, Ferrari AC, Geim AK et al (2009) Subjecting a graphene monolayer to tension and compression. Small 5(21):2397–2402
45. Palermo V, Kinloch IA, Ligi S, Pugno NM (2016) Nanoscale mechanics of graphene and graphene oxide in composites: a scientific and technological perspective. Adv Mater 28(29):6232–6238
46. Raju APA, Lewis A, Derby B, Young RJ, Kinloch IA, Zan R et al (2014) Wide-area strain sensors based upon graphene-polymer composite coatings probed by Raman spectroscopy. Adv Funct Mater 24(19):2865–2874

47. Anagnostopoulos G, Pappas PN, Li Z, Kinloch IA, Young RJ, Novoselov KS et al (2016) Mechanical stability of flexible graphene-based displays. ACS Appl Mater Interfaces 8(34):22605–22614
48. Wang Y, Yang R, Shi Z, Zhang L, Shi D, Wang E, Zhang G (2011) Super-elastic graphene ripples for flexible strain sensors. ACS Nano 5(5):3645–3650
49. Frank O, Tsoukleri G, Parthenios J, Papagelis K, Riaz I, Jalil R et al (2010) Compression behavior of single-layer graphenes. ACS Nano 4(6):3131–3138
50. Nayyar V, Ravi-Chandar K, Huang R (2014) Stretch-induced wrinkling of polyethylene thin sheets: experiments and modeling. Int J Solids Struct 51(9):1847–1858
51. Yang K, Chen Y, Pan F, Wang S, Ma Y, Liu Q (2016) Buckling behavior of substrate supported graphene sheets. Materials 9(1):32
52. Zhao Q, Qi HJ, Xie T (2015) Recent progress in shape memory polymer: new behavior, enabling materials, and mechanistic understanding. Prog Polym Sci 49:79–120
53. Hager MD, Bode S, Weber C, Schubert US (2015) Shape memory polymers: past, present and future developments. Prog Polym Sci 49:3–33
54. Zhang RF, Ning ZY, Zhang YY, Zheng QS, Chen Q, Xie HH et al (2013) Superlubricity in centimetres-long double-walled carbon nanotubes under ambient conditions. Nat Nanotechnol 8(12):912–916
55. Liu Y, Xie B, Xu Z (2011) Mechanics of coordinative crosslinks in graphene nanocomposites: a first-principles study. J Mater Chem 21(18):6707–6712
56. Criegee R (1975) Mechanism of ozonolysis. Angew Chem Int Ed 14(11):745–752
57. Ramanathan T, Abdala AA, Stankovich S, Dikin DA, Herrera-Alonso M, Piner RD et al (2008) Functionalized graphene sheets for polymer nanocomposites. Nat Nanotechnol 3(6):327–331
58. Putz KW, Compton OC, Palmeri MJ, Nguyen ST, Brinson LC (2010) High-nanofiller-content graphene oxide-polymer nanocomposites via vacuum-assisted self-assembly. Adv Funct Mater 20(19):3322–3329
59. Nemanich RJ, Lucovsky G, Solin S (1977) Infrared active optical vibrations of graphite. Solid State Commun 23(2):117–120
60. Niyogi S, Bekyarova E, Itkis ME, Zhang H, Shepperd K, Hicks J et al (2010) Spectroscopy of covalently functionalized graphene. Nano Lett 10(10):4061–4066
61. Engiert JM, Vecera P, Knirsch KC, Schafer RA, Hauke F, Hirsch A (2013) Scanning-Raman-microscopy for the statistical analysis of covalently functionalized graphene. ACS Nano 7(6):5472–5482
62. Wang MC, Lai ZB, Galpaya D, Yan C, Hu N, Zhou LM (2014) Atomistic simulation of surface functionalization on the interfacial properties of graphene-polymer nanocomposites. J Appl Phys 115(12):123520
63. Brennan CJ, Nguyen J, Yu ET, Lu NS (2015) Interface adhesion between 2D materials and elastomers measured by buckle delaminations. Adv Mater Interfaces 2(16):1500176
64. Jiang H, Khang DY, Song J, Sun Y, Huang Y, Rogers JA (2007) Finite deformation mechanics in buckled thin films on compliant supports. Proc Natl Acad Sci 104(40):15607–15612
65. Faulhaber S, Mercer C, Moon M, Hutchinson J, Evans A (2006) Buckling delamination in compressed multilayers on curved substrates with accompanying ridge cracks. J Mech Phys Solids 54(5):1004–1028
66. Zhang YY, Wang CM, Cheng Y, Xiang Y (2011) Mechanical properties of bilayer graphene sheets coupled by sp^3 bonding. Carbon 49(13):4511–4517
67. Schniepp HC, Li JL, McAllister MJ, Sai H, Herrera-Alonso M, Adamson DH et al (2006) Functionalized single graphene sheets derived from splitting graphite oxide. J Phys Chem B 110(17):8535–8539
68. Gómez-Navarro C, Burghard M, Kern K (2008) Elastic properties of chemically derived single graphene sheets. Nano Lett 8(7):2045–2049
69. Zandiatashbar A, Lee GH, An SJ, Lee S, Mathew N, Terrones M et al (2014) Effect of defects on the intrinsic strength and stiffness of graphene. Nat Commun 5:3186
70. Liu P, Jin Z, Katsukis G, Drahushuk LW, Shimizu S, Shih CJ et al (2016) Layered and scrolled nanocomposites with aligned semi-infinite graphene inclusions at the platelet limit. Science 353(6927):364–367

71. Shang J, Chen Y, Zhou Y, Liu L, Wang G, Li X et al (2015) Effect of folded and crumpled morphologies of graphene oxide platelets on the mechanical performances of polymer nanocomposites. Polymer 68:131–139
72. Young RJ, Liu M, Kinloch IA, Li S, Zhao X, Vallés C et al (2018) The mechanics of reinforcement of polymers by graphene nanoplatelets. Compos Sci Technol 154:110–116
73. Papageorgiou DG, Li Z, Liu M, Kinloch IA, Young RJ (2020) Mechanisms of mechanical reinforcement by graphene and carbon nanotubes in polymer nanocomposites. Nanoscale 12(4):2228–2267
74. Zhao C, Zhang P, Zhou J, Qi S, Yamauchi Y, Shi R et al (2020) Layered nanocomposites by shear-flow-induced alignment of nanosheets. Nature 580(7802):210–215
75. Pang Y, Yang J, Curtis TE, Luo S, Huang D, Feng Z et al (2019) Exfoliated graphene leads to exceptional mechanical properties of polymer composite films. ACS Nano 13(2):1097–1106

Chapter 3
Mechanical Behavior at Graphene/Polymer Interfaces Under Biaxial Compression

3.1 Introduction

The discovery of graphene brought an immense boost up and new dimension to modern electronic devices due to outstanding properties, such as the high carrier mobility, transparency, flexibility, conductivity and mechanical strength [1–3]. In particular, flexible electronics stand to benefit from the development of graphene because their unmatched combination of device physics and device mechanics is accessible on soft polymeric substrates [4–7]. Many research groups have demonstrated the resilience of graphene-based flexible electrodes as a promising candidate in a wide range of applications, such as field effect transistors (FETs), solar cells, organic light-emitting diodes (OLEDs), touch screens, and electrochemical sensors [8–12]. Apart from the high electrical, optical performance as required by flexible opto-electronic devices, the mechanical stability and recoverability of electrodes under deformation were major factors to determine the durability and lifetime of devices [13]. Various mechanical deformations such as tension, compression, bending, and twisting would apply to the device under the various harsh working conditions during the fabrication, assembling, and manipulation processes [14–18]. The occurrence of mechanical failure could severely deteriorate the performance and reliability when the desired properties of the flexible devices are proposed to be tunable through reversible and repeatable mechanical manipulations [19–21]. Therefore, it is crucially important to monitor the mechanical responses of flexible electrodes under various stress states, evaluate the interface stability and recoverability, clarify the possible failure modes as well as the limit of strain-induced irreversible structure deformation.

The ultralow bending rigidity enables graphene to easily out-of-plane deform when subjected to compression loads [22–24]. The consequently generated wrinkles and buckles greatly threaten the mechanical stability of graphene-based flexible electrodes. To this end, substantial efforts are devoted to exploring the mechanical behavior of graphene sandwiched between polymers layers in response to compressive stress. Galiotis et al. investigated in detail both stress uptake and buckling strain

G. Wang, *Characterization and Modification of Graphene-Based Interfacial Mechanical Behavior*, Springer Theses, https://doi.org/10.1007/978-981-15-8029-1_3

of embedded graphene under compression using the polymer cantilever beam technique. The critical buckling strain was found to be greatly dependent on the length-to-width ratios (l/w) of the graphene flake, which was suppressed by the physical constraints of surrounding polymer matrix [25]. In addition, theoretical simulation based on coarse-grain model has predicted the buckling deformation of the embedded graphene under compression, and revealed a dependency of deformation mode on the adhesion strength and effective stiffness of graphene and substrate [26]. It should be noted, however, that the deformation modes aforementioned are relatively simple and fail to be comparable to the stress states of flexible electrodes suffered in reality. In general, for thin solid film-based flexible electrodes, the stress states within the embedded graphene sheets come from at least biaxial directions. Recently, Zang et al. realized the biaxial compression of graphene adhered on polydimethylsiloxane (PDMS), by relaxing the pre-stretched substrate sequentially along two orthogonal directions [27]. It was observed that the graphene film develops wrinkles and buckle delamination when the substrate is relaxed uniaxially, and further becomes crumpled when the substrate is relaxed biaxially. However, a deeper understanding of the mechanical mechanism underlying different buckling modes and mechanical probing of interfacial parameters are still lacking [28, 29].

To reveal the mechanical response and evaluate mechanical stability of the embedded graphene, in this chapter, we systematically investigated the mechanical deformation of graphene incorporated in PMMA when subjected to biaxial compression. In terms of the thermal expansion mismatch between graphene and PMMA, liquid nitrogen was employed to induce a thermal shrinkage. The compressive deformation modes as well as the microstructural evolution of the embedded graphene sheets were monitored by in situ micro-Raman spectroscopy. A simplified model was further proposed to extract key parameters and unveil the failure mechanism at the microscale [30].

3.2 Experimental Methods

3.2.1 Preparation of Graphene/Polymer Sandwich Structure

Polymethyl methacrylate (PMMA) powders were placed between two steel plates, which were then pressed under the hot-pressing machine (KH-0801) for 15 min at a temperature of 200 °C. After cooling to room temperature, the film was removed from the mold and the thickness was measured to be around 200 μm. Following the wet transfer method as introduced in Sect. 2.2.1, we transferred the mechanically exfoliated monolayer graphene onto the PMMA substrate, keeping the capping layer (~100 nm) on top to fabricate the sandwich structure. The layer number of graphene was identified by optical microscopy and Raman spectroscopy, as shown in Fig. 3.1. Apparently, the different contrast regions in Fig. 3.1a are related to flakes with different numbers of graphene layers, which are consistent with the contour

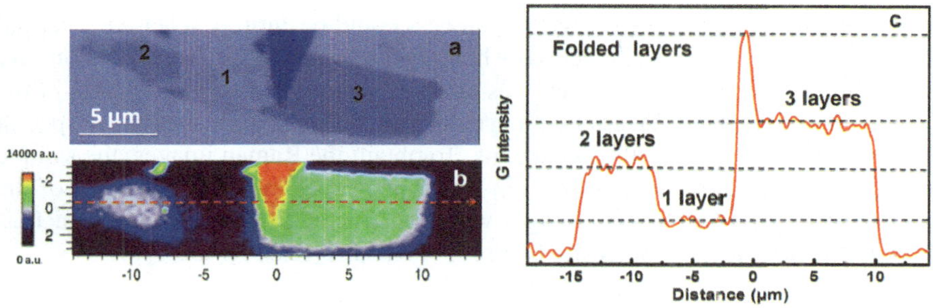

Fig. 3.1 **a** Optical micrograph of graphene flakes containing monolayer, bilayer, and trilayer regions. **b** Raman image of graphene flakes plotted by the intensity of Raman G-band. **c** The intensity cross-section of the Raman image corresponding to the dash line in **b**

image of Raman G-band. It is clear that the stronger Raman G-band intensity corresponds to higher thickness of graphene layers [31, 32]. Here, we used a Renishaw Raman spectrometer equipped with the 514.5 nm line of an Ar laser. The laser power was kept below 0.5 mW on the sample to avoid laser-induced local heating of the sample. The incident light was polarized along the strain direction. The focused laser spot was approximately 1 μm in diameter, and the spectral resolution was $\sim$1 cm^{-1}.

3.2.2 *In-Plane Biaxial Compression via Thermal Shrinkage*

Graphene samples were placed into a Linkam cooling cell (TMS94) and temperature was precisely controlled under then Raman spectroscope, as shown in Fig. 3.2. The temperatures in our experiments ran from 60 down to -150 °C with a cooling rate of 10 °C/min by liquid nitrogen as the cooling agent. The liquid helium was further employed to introduce larger compressive strains to the embedded graphene when the ambient temperature was reduced to -263 °C (10 K). Here the temperature at 60 °C was selected as a reference based on the fact that the thermal stress of embedded

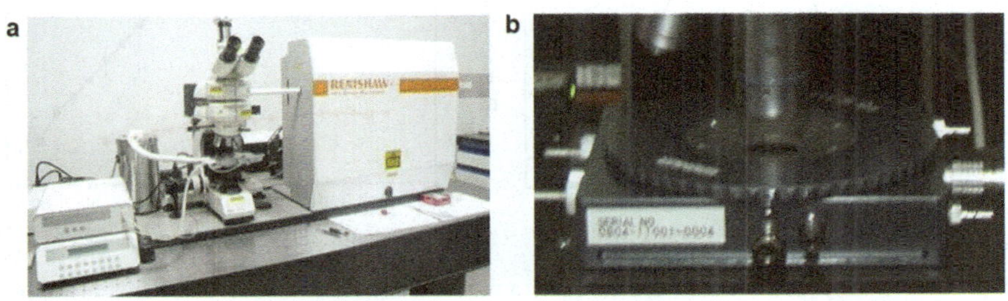

Fig. 3.2 **a** Raman spectroscopy combined with temperature controlling system. **b** Temperature controlled hot-stage for in-situ Raman measurement

graphene maintained during the curing process could be entirely released. Thermal equilibrium in each case was reached by holding the specimen for 3–5 min at constant temperature. The Raman spectrum was taken at a minimum of three points in the central area of graphene flake at a certain temperature. All bands in the Raman spectra of graphene were fitted with Lorentzians. To obtain the Raman images, the sample was moved with a step size of 300 nm and Raman spectra from every spot of the sample were recorded. Raman images were constructed using a parameter such as peak frequency and peak intensity by using WIRE Project software.

3.3 Biaxial Compression Graphene Embedded in Polymer Matrix

3.3.1 In Situ Raman Measurement

In general, thermal stress would be introduced to a material to constrain against expansion or contraction as the environmental temperature changes [33]. For PMMA/graphene/PMMA systems, the thermal stress acting on graphene comes from the differences in coefficients of thermal expansion (CTE) between graphene sheets and PMMA matrix. Different from stress state of embedded carbon nanotube (CNT) fiber reported earlier [33], in which the thermal stress mainly caused the compression along the fiber axis during the cooling process, a biaxial state of compressive stress was applied to the embedded graphene as the temperature decreased. Accordingly, the upward shifts of Raman G-band can be observed in Fig. 3.3a, due to the shortening

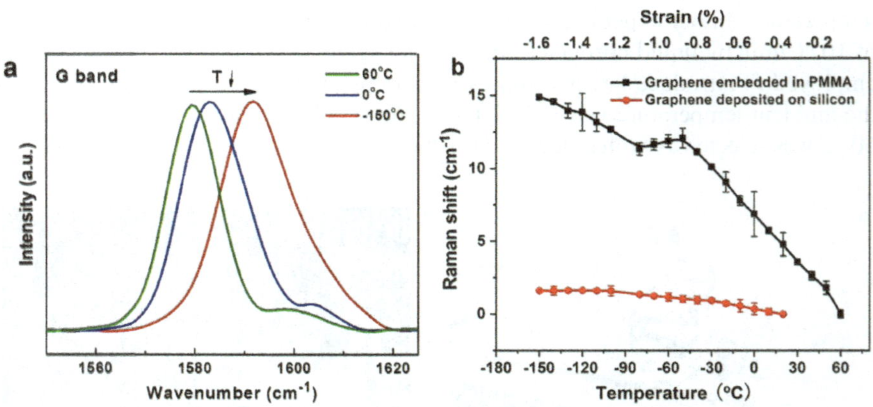

Fig. 3.3 **a** Variations of Raman G-band spectra of the embedded graphene monolayer at various temperatures. **b** The shift of Raman G-band as a function of applied temperatures (referred 60 °C) and corresponding compressive strain, for graphene embedded in PMMA and deposited on silicon, respectively

of the C=C bond length of graphene. Meanwhile, the peak broadening is also visible upon decreasing temperature, implying a non-uniform strain distribution within the graphene flake. To exclude the thermal effect on the Raman shifts, we performed the same cooling procedures for graphene deposited on silicon wafer. As presented in Fig. 3.3b, when the temperature decreases from 60 to −15 °C, a of 15 cm^{-1} was Raman G-band shows a negligible blueshift (2 cm^{-1}), in striking contrast to that of embedded graphene (15 cm^{-1}). Therefore, it is the strain effect rather than the thermal effect that dominates the Raman shift, in consistency with previous reports [34].

Accompanying the continuous reduction of temperature, the shifts of Raman G-band peak position exhibit three-stage features and delineate the microstructural deformation, as shown in Fig. 3.3b. In the initial stage, the graphene flake suffers from the elastic compression. The stress applied to the PMMA matrix is efficiently transferred to graphene across the interface, rendering the Raman G-band increase linearly with decreasing temperature; Once the temperature exceeds a certain point (~−50 °C), the Raman G-band reach a plateau region. This is associated with the onset of Euler buckling to release the compressive stress in graphene; after that, Raman G-band appears a sluggish up-shifted tendency with further decrements of temperature down to −150 °C, suggesting that the graphene sheets could be continuously compressed with further increasing compressive strain. Through monitoring the compressive deformation of the embedded graphene, the load-bearing status, deformation mechanism, its mechanical stability and recoverability of the embedded graphene can be derived as follows.

3.3.2 *Elastic Compression*

The magnitude of thermal stress in graphene is influenced by the differences in the elastic modulus between the two components and the temperature gradient of the systems. During the cooling process, the compressive strain (ε) for the PMMA/graphene/PMMA systems can be inferred from the following equation according to the classical elasticity theory [35]:

$$\varepsilon = (\alpha_{PMMA} - \alpha_{graphene})\Delta T \tag{3.1}$$

where ΔT is the temperature gradient with 60 °C as the reference temperature, $\alpha_{PMMA} = (68.0 \pm 2.0) \times 10^{-6}\ \mathrm{K}^{-1}$ and $\alpha_{graphene} = (-8.0 \pm 0.7) \times 10^{-6}\ \mathrm{K}^{-1}$ are the coefficients of thermal expansion (CTE) of the PMMA matrix and graphene, respectively [36, 37]. Thus, the quantitative relationship between macroscopic applied strain and Raman shift can be deduced as shown in Fig. 3.3b. In the elastic compression stage, up to approximately −0.6% strain, the graphene sheet exhibits a reversible compressive behavior. The shift of Raman G-band increased linearly with applied compressive strain with shift rate ($\partial\omega_G/\partial\varepsilon$) in the range of ~12.1 to ~15.6 cm^{-1}/%, and no apparent G-band splitting was observed due to the biaxial loading. The scattered results in shift rate could be attributed to the variation in flake sizes as well as residual strain. Compared to the shift rates of ~10.8 and ~31.7 $cm^{-1}/\%$ for the G^+ and G^- mode under uniaxial compression as previously reported [25], the apparent difference could be related to the different compression modes (uniaxial vs. biaxial), varied lateral dimensions of graphene flakes and different interfacial adhesion [38].

In addition, the local stress can be estimated according to the Raman shift of graphene under compression. In the presence of strain, the dynamical equations that describe phonon modes in the solid have the following form [39, 40]:

$$\bar{m}\ddot{u}_i = -\sum_j K_{ij}u_j = -\left(K_{ii}u_i + \sum_{jkl}\frac{\partial K_{ij}}{\partial \varepsilon_{kl}}\varepsilon_{kl}u_j\right) \tag{3.2}$$

where u_i is the ith component of the relative displacement of two atoms in one unit cell, $\bar{m}$ is the reduced mass of two atoms, $K_{ij} = \bar{m}\omega_0^2$ is the effective spring constant of the phonon modes in the absence of strain. The double-dotted quantities indicate second derivatives. $\frac{\partial K_{ij}}{\partial \varepsilon_{kl}}\varepsilon_{kl}$ is the change in the spring constant due to the applied strain (ε_{kl}) while i, j, k, and l denote crystallographic axes (Fig. 3.4a). Taking into account the hexagonal symmetry of the graphene lattice and the absence of shear components of strain, the result for the Raman active E_{2g} mode is given by

$$\begin{vmatrix} A(\varepsilon_{xx} + \varepsilon_{yy}) - \lambda & B(\varepsilon_{xx} - \varepsilon_{yy}) \\ B(\varepsilon_{xx} - \varepsilon_{yy}) & A(\varepsilon_{xx} + \varepsilon_{yy}) - \lambda \end{vmatrix} = 0 \tag{3.3}$$

where ε_{xx} and ε_{yy} are the strains along x- and y-directions respectively, $\lambda = \omega_\sigma^2 - \omega_0^2 \approx 2\omega_0(\omega_\sigma - \omega_0)$, ω_σ is the strain-dependent Raman frequency and ω_0 is the frequency in the absence of strain. There is only one solution for this secular equation:

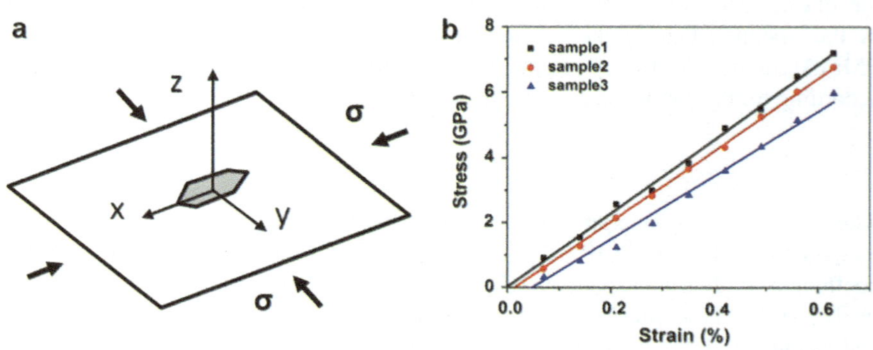

Fig. 3.4 **a** Coordinate system for a crystallite in graphene. **b** Typical stress–strain curves for monolayer graphene embedded in PMMA

$$\lambda = A(\varepsilon_{xx} + \varepsilon_{yy}) \quad (3.4)$$

On the other hand, for hexagonal crystals, symmetry reduces the generalization of Hooke's law which relates the biaxial stress to the strain as follows [41]:

$$\begin{bmatrix} \varepsilon_{xx} \\ \varepsilon_{yy} \\ \varepsilon_{zz} \\ \varepsilon_{yz} \\ \varepsilon_{zx} \\ \varepsilon_{xy} \end{bmatrix} = \begin{bmatrix} S_{11} & S_{12} & S_{13} & & & \\ S_{12} & S_{11} & S_{13} & & & \\ S_{13} & S_{13} & S_{33} & & & \\ & & & S_{44} & & \\ & & & & S_{44} & \\ & & & & & 2(S_{11} - S_{12}) \end{bmatrix} \begin{bmatrix} \sigma \\ \sigma \\ 0 \\ 0 \\ 0 \\ 0 \end{bmatrix} \quad (3.5)$$

with coordinate x and y in the graphene basal plane and z out of the plane, so that $\varepsilon_{xx} = \varepsilon_{yy} = (S_{11} + S_{12})\sigma$, $\varepsilon_{zz} = 2S_{13}\sigma$, $\varepsilon_{yz} = \varepsilon_{zx} = \varepsilon_{xy} = 0$, where S_{ij} is the compliance coefficient. We hence obtain $\varepsilon_{xx} = \varepsilon_{yy} = (S_{11} + S_{12})\sigma$, and combine with Eq. (3.4), the stress is expressed as:

$$\sigma = \frac{\omega_0(\omega_\sigma - \omega_0)}{A(S_{11} + S_{22})} \quad (3.6)$$

where $A = \omega_0 \frac{\partial \omega_\sigma}{\partial \varepsilon} = -2.47 \times 10^6\ \text{cm}^{-1}$ is the phenomenological coefficient and $\omega_0 = 1580\ \text{cm}^{-1}$. Using graphite elastic constant $S_{11} = 0.98 \times 10^{-12}\ \text{Pa}^{-1}$ and $S_{12} = -0.16 \times 10^{-12}\ \text{Pa}^{-1}$, the stress–strain relationship is obtained as shown in Fig. 3.4b [42]. In particular, the compressive stress acting on the graphene monolayer could reach as high as ~6 GPa at the strain level of ~−0.6%. Such an extraordinary load-bearing capability of monolayer graphene sheet plays a significant role in the mechanical stability and creep-resistance of graphene-based flexible electronics in practical application.

3.3.3 *Euler Buckling Instability*

Due to the atomic-level thickness and high flexibility, graphene will buckle instead of fracture when the strain is increased to a high level. The stress relaxation hence causes the Raman G-band to reach a plateau region at a critical strain of around −0.8%, as shown in Fig. 3.5a. On the basis of mechanical instability theory, the critical buckling strain ($\varepsilon_{buckling}$) for a rectangular thin shell under biaxial compression is determined by [43]:

$$\varepsilon_{buckling} = k\frac{D\pi^2}{C} \quad (3.7)$$

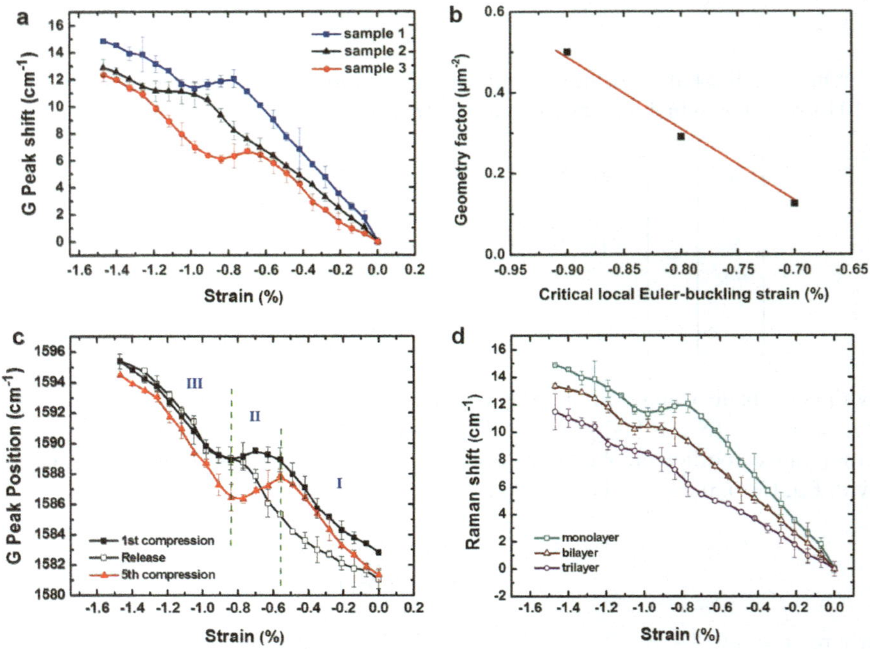

Fig. 3.5 **a** The shift of Raman G-band as a function of compressive strain for three graphene flakes with different sizes. **b** Geometric term *k* plotted against critical local Euler buckling strain ($\varepsilon_{buckling}$) for three graphene flakes under study. The line represents a linear fit to the experimental points. **c** Raman G-band position versus strain over a compression/relaxation cycle. **d** The shift of Raman G-band as a function of compressive strain for graphene flakes with layer numbers

where D and C are the flexural and tension rigidities of materials, respectively, and k is the geometric term that lies in the size dimension combined with the numbers of half-waves where the flake buckles along the strain axis, as represented by [43]:

$$k = \frac{m^2}{l^2} + \frac{n^2}{w^2} \tag{3.8}$$

where l is the length and w is the width of graphene sheet, m is the number of half-waves in the length direction and n is the number in the width direction. Typically, at the ratio of $l/w = \sqrt{m(m+1)}$ we have the transition from m to $(m + 1)$ half-waves. For example, for the graphene flake with length $l = 4$ μm and width $w = 2$ μm, two and one half-waves are expected to occur along length and width directions, respectively, and thus $k = 0.5$. Herein, we compared the compressive behaviors of three individual graphene monolayers with different lateral dimensions. As seen, the three-stage compressive behaviors were observed for all testing samples (Fig. 3.5a). If we plot k as a function of $\varepsilon_{buckling}$, a linear relation is obtained as shown in Fig. 3.5b, in which $\varepsilon_{buckling}$ reduces as k falls off. It is hence concluded that even with the equal l/w value, the variation in lateral dimensions will lead to distinct size effects. Furthermore, based on the linear fitting by Eq. (3.7), the apparent bending rigidity of embedded graphene is estimated to be 8.4 MPa μm^3, which is 6 orders of magnitude higher than that of suspended graphene [38]. The considerably high flexural rigidity demonstrates that surrounding polymer matrix could provide a dramatic improvement to its compression behavior and it has an inspired significance to improve the stability of flexible electrodes.

It is noted that, the graphene sheets could be continuously compressed beyond the plateau stage with further increasing compressive strain, as evidenced by a sluggish up-shifted trend (~-10.4 cm^{-1}/%) of Raman G-band position (Fig. 3.5a). This unusual phenomenon implies that the graphene sheet in our experiments could continuously carry the external loads without permanent failure, whereas the irreversible collapse at interfacial regions occurred under uniaxial compression in literature [25, 38]. We suggest that the excellent compressibility of the PMMA/graphene/PMMA system at large applied compressive strain could be attributed to the excellent interfacial stability and the different deformation modes. Tsoukleri et al. reported the occurrence of the irreversible interface failure occurred as the compressive strain was up to ~0.7% for the PMMA/graphene/SU8 photoresist system [38]. The brittle feature of the SU8 photoresist resulting from its intrinsic three-dimensional network-like structure would greatly inhibit the interface stability of graphene-based electrodes at large applied compressive strain, in which the cracks easily formed at the SU8 layer. Instead, PMMA with linear structure feature behaved excellent plastic characteristics under deformation, and thereby no apparent interfacial failure was observed in our system.

In order to confirm the well-bonded interface between the graphene and PMMA, we further conduct the cyclic compression/relaxation tests through controlling temperature procedures as illustrated in Fig. 3.5c. In the first-cycle compressive deformation with strain up to -1.5%, the mechanical behavior of embedded graphene platelet was similar to that shown in Fig. 3.3b. After that, with the increase of temperature accompanying by the relaxation of both matrix and graphene sheet, the Raman G-band shows apparent down-ward shifts, indicating the relaxation of the compressed C=C bond of graphene. Moreover, the shifted trend of the Raman G-band along unloading part was slightly different from that along the loading one. Specifically, no apparent plateau region was observed during the unloading process due to the slow recovery of the buckled graphene sheets. But the shifted rates of the Raman G-band for Stage-I and Stage-III were kept consistent during the loading/unloading process, with the slopes ~-12.2 and ~-11.9 cm^{-1}/% in loading and ~-13.1 and ~-12.4 cm^{-1}/% in unloading, respectively. After five-cycle loading/unloading treatment, the shifted trends of the Raman G-band in Stage-I and Stage-III still keep the similar constants as the initial one, implying the good recoverability of the embedded graphene sheet. But it is worth noting that the initiation of local Euler-type buckling slightly shifted to low compressive strain level due to the repetitive loading/unloading treatment induced weaker interfacial adhesion.

We further investigate the strain-sensitive Raman shift for few-layer graphene under compression, as shown in Fig. 3.5d. Apparently, as the layer number increases, the Raman shift rate in the elastic compression stage decreases, while the critical buckling strain increases. The plateau behavior almost vanishes for trilayer graphene. This can be explicable based on the stress transfer mechanism. In few-layer graphene, additional interlayer interfaces are introduced, which is demonstrated to be much weaker than graphene/polymer interface [44]. Thus, the overall mechanical response is lowered by the poor efficiency of stress transfer between graphene layers. On the other hand, such a weak interaction easily induces the relative sliding between these atomically smooth layers [45], which will further hinder the load bearing and reduce the Raman shift of strained graphene.

3.3.4 Local Debonding

To further explore the deformation process and clarify the possible fracture mechanism of the embedded graphene sheets, herein, we increased the compressive strain level up to ~2.3% by using liquid He as cooling agent. Given that strain-induced shift of the Raman 2D-band of monolayer graphene platelet is twice larger than that of the Raman G-band under tensile/compression deformation [46], therefore, we utilized Raman 2D-band to reveal the strain distribution of the embedded graphene sheets before and after mechanical treatment. Figure 3.6a shows the Raman 2D-band image mapping of the monolayer graphene sheet experienced the apparently residual compressive stress resulting from the sample preparation process. As seen, the residual compressive strain of the embedded graphene sheet is roughly uniformly distributed within the whole area (sample size 3 μm × 5 μm) with slight variation as indicated by arrows. Considering that a piezo stage with a step size of ~300 nm

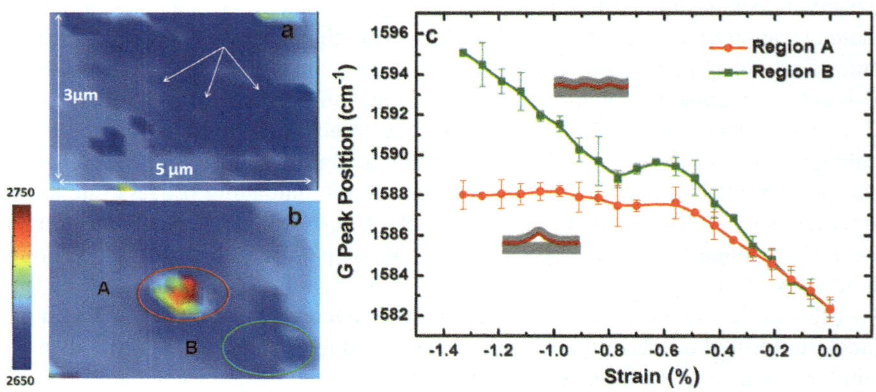

Fig. 3.6 Raman mapping image of monolayer graphene plotted by the peak position of Raman 2D-band **a** before and **b** after the large compressive deformation. **c** Peak position of Raman G-band as a function of strain for the two regions as presented by circles in **b**

and a Raman spectrum was recorded at every point, thus, we attributed the variation in strain distribution to the wrinkled feature of the graphene sheets during sample fabrication process [38, 47]. After mechanical loading/unloading treatment with maximum compressive strain up to ~2.3%, we could see that the strain distribution of the graphene sheet turned highly non-uniform as shown in Fig. 3.5b, indicated by solid-line circles. The center part in the sheet suffered apparently residual compressive strain as compared with its neighboring parts. To confirm the occurrence of permanent deformation at the interfacial region under large applied compressive strain levels, herein, we further applied biaxial compression to the sample and investigated the shift rate of Raman G-band in specific Region A and Region B. Apparently, in Fig. 3.6c, the Region B displays typical three-stage features with shift rate of $-13.3\ cm^{-1}/\%$ (Stage-I), $-12.1\ cm^{-1}/\%$ (Stage-III) respectively. The shifted rates are close to that in Fig. 3.5c, further indicating the excellent recoverability of the localized graphene sheets under mechanical deformation. In stark contrast, the shifted trend of Raman G-band in Region A is quite different from that in Region B, where the shift rate (Stage-I) decreased to $-9.1\ cm^{-1}/\%$ as compared to that in Region B ($-13.3\ cm^{-1}/\%$), and no further G-band shift was observed after reaching plateau region (~-0.6%). We hence speculate the ignition of interfacial debonding once the strain level reached to ~-2.3%. After that, the strong interfacial adhesion turns weak. Consequently, the weaken interfacial adhesion in Region A leads to poor interfacial stress transfer efficiency, and eventually the interfacial debonding occurred at relative low strain level in the PMMA/graphene/PMMA system. As expected, such areas would act as stress concentration sites to induce interfacial delamination when subjected to further mechanical manipulation.

Our observation is also reminiscent of the buckling mode selection map proposed by Mei and Huang [48, 49]: The transition between the wrinkling and buckle delamination is predicted as the stiffness ratio between the substrate and the film or the interfacial defect size varies. Note that, the wrinkling mode actually corresponds to Euler buckling here, requiring the substrate to deform coherently with the film; the buckle delamination is actually the debonding failure that is typically localized and sensitive to interfacial defects. The schematic illustration of two different buckling modes are depicted in the insets of Fig. 3.6c. Given the significant stiffness mismatch between graphene and PMMA ($E_{graphene}/E_{PMMA} \sim 10^3$), periodic wrinkling is energetically more favorable than buckle delamination in the current work. Therefore, the critical buckling strain is first reached during the compression, beyond which the conformal adherence between graphene and PMMA maintains the effective interfacial stress transfer and allows for the mechanical recoverability. As the temperature is further decreased, the compressive stress eventually reaches the critical debonding stress for the initial defect size, and buckle delamination occurs alongside the existing wrinkles. Consequently, the interfacial defect size increases due to the growth of buckle delamination, thus lowering the critical debonding strain compared to the first cooling. As seen in Fig. 3.6c, the buckle delamination occurred prior to wrinkling in the second cycle, leading to the failure in the stress transfer.

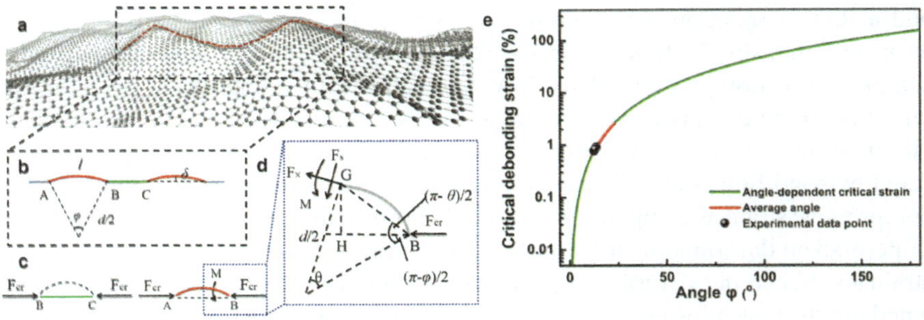

Fig. 3.7 **a** Wrinkled graphene sheet by molecular dynamics simulation. **b** Two-dimensional graphene sheet was simplified as one-dimensional bent beam which consists of **c** the weak interface part BC and the initial wrinkled part AB with radius $d/2$ and central angle φ. **d** Schematic of force analysis on the part AB. **e** Plot of central angle (φ) of wrinkled graphene sheets versus critical debonding strain

To deeply understand the interfacial debonding behavior in typical graphene-polymer systems as presented in Fig. 3.6b, we proposed a theoretical model by simplifying two-dimensional graphene sheet as a one-dimensional (1D) elastic column based on the mechanical instability theory as depicted in Fig. 3.7a. Considering the wrinkled nature of suspended graphene sheet, the embedded graphene sheet would still maintain such wrinkled morphology, though suppressed, within polymeric matrix [50, 51]. Herein, we presumed that part of the embedded graphene sheet sustains the initially wrinkled morphology with relatively strong interfacial interaction as identified by AB part in Fig. 3.7b, while the other part exhibit flat morphology with relative weak interfacial adhesion identified by BC part. As expected, when subjected to mechanical manipulation, the interfacial debonding would initiate at areas with weak adhesion and then propagate to the areas with strong adhesion, eventually leading to the interfacial delamination. Thus, the strain level applied to the graphene-polymer system to induce mechanical debonding is a critical factor to affect the mechanical stability of graphene-based flexible electrodes. Given the relatively weak bonding at BC part, only horizontal forces are considered here. We take a cross section for analysis and regard it as a 1D elastic column. According to the theory of elastic stability, the critical force (F_{cr}) for the buckling is

$$F_{cr} = \frac{\pi^2 EI}{l^2} \tag{3.9}$$

where EI is the bending rigidity, l is the length of BC part. Figure 3.7d shows the force analysis of AB part, where the bending moment is [35]:

$$M = F_{cr} \cdot GH = F_{cr} \cdot BG \cdot \sin\left(\frac{\pi - \theta}{2} - \frac{\pi - \varphi}{2}\right) = F_{cr} d \sin\frac{\theta}{2} \sin\frac{\varphi - \theta}{2} \tag{3.10}$$

where d and φ are the diameter and central angle of AB arc, θ is the central angle of infinitesimal element taken from AB arc. The elastic strain energy of AB part is given as:

$$V_{\varepsilon} = \int_0^{\varphi} dV_{\varepsilon} = \int_0^{\varphi} \frac{M^2(d/2)}{2EI} d\theta = \frac{F_{cr}^2 d^3}{4EI}\left(\frac{1}{4}\varphi\cos^2\frac{\varphi}{2} + \frac{\varphi}{8} - \frac{3}{8}\sin\varphi\right) \tag{3.11}$$

According to the energy method, we have

$$V_{\varepsilon} = \frac{1}{2}F_{cr}\delta \tag{3.12}$$

Combining Eqs. (3.9), (3.11) and (3.12), the critical strain ($\varepsilon_{debonding}$) to predicate the interfacial debonding at the PMMA/graphene interface with weak adhesion under compression is represented as

$$\varepsilon_{debonding} = \frac{\pi^2}{\varphi^2\sin\frac{\varphi}{2}}\left(\varphi\cos^2\frac{\varphi}{2} + \frac{\varphi}{2} - \frac{3}{2}\sin\varphi\right) \tag{3.13}$$

Specially, the values of φ range from 9.3° to 22.8° in the reference [52, 53], and thus the calculated $\varepsilon_{debonding}$ is in the range of −0.4% to −2.2%, in which the large value implies strong interfacial adhesion while the low value corresponds to poor interfacial adhesion. Experimentally, the derived $\varepsilon_{debonding}$ is well consistent with the mechanical model as plotted in Fig. 3.7e, implying that the interfacial debonding initiated from the weak interface region. Therefore, the critical debonding strain is an important parameter to guide the mechanical manipulation of graphene-based flexible electrodes with various interfacial adhesion.

3.4 Chapter Summary

With the help of in situ micro-Raman spectroscopy, we systematically investigate the mechanical properties and its microstructure evolution of the individual graphene sheet inside flexible PMMA matrix when subjected to biaxial compression deformation. Our results demonstrated that the graphene could continuously carry compressive loads without permanent failure even after critical local Euler-type buckling strain (~−0.8%). The graphene/PMMA interface is able to survive with outstanding mechanical recoverability after cyclic loading at relative low strain level. However, with continuously increasing compressive strain up to ~−2.3%, the interfacial areas turned weak, and eventually, behave as the interfacial defects to accelerate debonding. Beyond that, combining the excellent effective flexural rigidity and remarkable compressive stress of the compressed graphene sheet, our work demonstrated the excellent reinforcing role of graphene played in the flexible electronics. It will be helpful not only to guide the design of graphene-based flexible electronics with excellent mechanical stability and durability, but also to deeply understand the deformation modes of various nanofillers based flexible electronic devices.

References

1. Geim AK, Novoselov KS (2007) The rise of graphene. Nat Mater 6(3):183–191
2. Geim AK (2009) Graphene: status and prospects. Science 324(5934):1530–1534
3. Nair RR, Blake P, Grigorenko AN, Novoselov KS, Booth TJ, Stauber T et al (2008) Fine structure constant defines visual transparency of graphene. Science 320(5881):1308–1308
4. Akinwande D, Petrone N, Hone J (2014) Two-dimensional flexible nanoelectronics. Nat Commun 5:5678–5689
5. Gomez De Arco L, Zhang Y, Schlenker CW, Ryu K, Thompson ME, Zhou C (2010) Continuous, highly flexible, and transparent graphene films by chemical vapor deposition for organic photovoltaics. ACS Nano 4(5):2865–2873
6. Das S, Gulotty R, Sumant AV, Roelofs A (2014) All two-dimensional, flexible, transparent, and thinnest thin film transistor. Nano Lett 14(5):2861–2866
7. Lee J, Ha T-J, Li H, Parrish KN, Holt M, Dodabalapur A et al (2013) 25 GHz embedded-gate graphene transistors with high-K dielectrics on extremely flexible plastic sheets. ACS Nano 7(9):7744–7750
8. Zhu Y, Sun Z, Yan Z, Jin Z, Tour JM (2011) Rational design of hybrid graphene films for high-performance transparent electrodes. ACS Nano 5(8):6472–6479
9. Choi W, Lahiri I, Seelaboyina R, Kang YS (2010) Synthesis of graphene and its applications: a review. Crit Rev Solid State Mater Sci 35(1):52–71
10. Wu J, Agrawal M, Becerril HA, Bao Z, Liu Z, Chen Y et al (2010) Organic light-emitting diodes on solution-processed graphene transparent electrodes. ACS Nano 4(1):43–48
11. Duan F, Li W, Wang G, Weng C, Jin H, Zhang H et al (2019) Can insulating graphene oxide contribute the enhanced conductivity and durability of silver nanowire coating? Nano Res 12(7):1571–1577
12. Ren H, Zheng L, Wang G, Gao X, Tan Z, Shan J et al (2019) Transfer-medium-free nanofiber-reinforced graphene film and applications in wearable transparent pressure sensors. ACS Nano 13(5):5541–5548
13. Harris K, Elias A, Chung H-J (2016) Flexible electronics under strain: a review of mechanical characterization and durability enhancement strategies. J Mater Sci 51(6):2771–2805
14. Li H, Al-Aqtash N, Wang L, Qin R, Liu Q, Zheng J et al (2012) Electromechanical switch in metallic graphene nanoribbons via twisting. Phys E Low Dimens Syst Nanostruct 44(10):2021–2026
15. Koskinen P (2012) Graphene nanoribbons subject to gentle bends. Phys Rev B 85(20):205429
16. Nathan A, Ahnood A, Cole MT, Lee S, Suzuki Y, Hiralal P et al (2012) Flexible electronics: the next ubiquitous platform. Proc IEEE 100:1486–1517
17. Dai Z, Weng C, Liu L, Hou Y, Zhao X, Kuang J et al (2016) Multifunctional polymer-based graphene foams with buckled structure and negative Poisson's ratio. Sci Rep 6:32989
18. Weng C, Dai Z, Wang G, Liu L, Zhang Z (2019) Elastomer-free, stretchable, and conformable silver nanowire conductors enabled by three-dimensional buckled microstructures. ACS Appl Mater Interfaces 11(6):6541–6549
19. Anagnostopoulos G, Pappas PN, Li Z, Kinloch IA, Young RJ, Novoselov KS et al (2016) Mechanical stability of flexible graphene-based displays. ACS Appl Mater Interfaces 8(34):22605–22614
20. Raju APA, Lewis A, Derby B, Young RJ, Kinloch IA, Zan R et al (2014) Wide-area strain sensors based upon graphene-polymer composite coatings probed by Raman spectroscopy. Adv Funct Mater 24(19):2865–2874

21. Chen H, Lu B-W, Lin Y, Feng X (2013) Interfacial failure in flexible electronic devices. IEEE Electron Device Lett 35(1):132–134
22. Tapasztó L, Dumitrică T, Kim SJ, Nemes-Incze P, Hwang C, Biró LP (2012) Breakdown of continuum mechanics for nanometre-wavelength rippling of graphene. Nat Phys 8(10):739–742
23. Wei Y, Wang B, Wu J, Yang R, Dunn ML (2012) Bending rigidity and Gaussian bending stiffness of single-layered graphene. Nano Lett 13(1):26–30
24. Lu Q, Arroyo M, Huang R (2009) Elastic bending modulus of monolayer graphene J Phys D Appl Phys 42(10):102002
25. Frank O, Tsoukleri G, Parthenios J, Papagelis K, Riaz I, Jalil R et al (2010) Compression behavior of single-layer graphenes. ACS Nano 4(6):3131–3138
26. Cranford SW (2013) Buckling induced delamination of graphene composites through hybrid molecular modeling. Appl Phys Lett 102(3):031902
27. Zang J, Ryu S, Pugno N, Wang Q, Tu Q, Buehler MJ et al (2013) Multifunctionality and control of the crumpling and unfolding of large-area graphene. Nat Mater 12(4):321–325
28. Ma L, He L, Ni Y (2020) Tunable hierarchical wrinkling: from models to applications. J Appl Phys 127(11):111101
29. Wang Q, Zhao X (2016) Beyond wrinkles: multimodal surface instabilities for multifunctional patterning. MRS Bull 41(02):115–122
30. Wang G, Liu L, Dai Z, Liu Q, Miao H, Zhang Z (2015) Biaxial compressive behavior of embedded monolayer graphene inside flexible poly (methyl methacrylate) matrix. Carbon 86:69–77
31. Li SL, Miyazaki H, Song H, Kuramochi H, Nakaharai S, Tsukagoshi K (2012) Quantitative Raman spectrum and reliable thickness identification for atomic layers on insulating substrates. ACS Nano 6(8):7381–7388
32. Li XL, Qiao XF, Han WP, Lu Y, Tan QH, Liu XL et al (2015) Layer number identification of intrinsic and defective multilayered graphenes up to 100 layers by the Raman mode intensity from substrates. Nanoscale 7:8135–8141
33. Gao Y, Li JZ, Liu LQ, Ma WJ, Zhou WY, Xie SS et al (2010) Axial compression of hierarchically structured carbon nanotube fiber embedded in epoxy. Adv Funct Mater 20(21):3797–3803
34. Calizo I, Balandin A, Bao W, Miao F, Lau C (2007) Temperature dependence of the Raman spectra of graphene and graphene multilayers. Nano Lett 7(9):2645–2649
35. Seely FB, Smith JO (1952) Advanced mechanics of materials. Wiley. New York
36. Yoon D, Son YW, Cheong H (2011) Negative thermal expansion coefficient of graphene measured by Raman spectroscopy. Nano Lett 11(8):3227–3231
37. Griffiths SK, Crowell JAW, Kistler BL, Dryden AS (2004) Dimensional errors in LIGA-produced metal structures due to thermal expansion and swelling of PMMA. J Micromech Microeng 14(11):1548–1557
38. Tsoukleri G, Parthenios J, Papagelis K, Jalil R, Ferrari AC, Geim AK et al (2009) Subjecting a graphene monolayer to tension and compression. Small 5(21):2397–2402
39. Sakata H, Dresselhaus G, Dresselhaus M, Endo M (1988) Effect of uniaxial stress on the Raman spectra of graphite fibers. J Appl Phys 63(8):2769–2772
40. Huang M, Yan H, Chen C, Song D, Heinz TF, Hone J (2009) Phonon softening and crystallographic orientation of strained graphene studied by Raman spectroscopy. Proc Natl Acad Sci 106(18):7304–7308
41. Ni Z, Chen W, Fan X, Kuo J, Yu T, Wee A et al (2008) Raman spectroscopy of epitaxial graphene on a SiC substrate. Phys Rev B 77(11):115416
42. Dresselhaus MS, Dresselhaus G, Sugihara K, Spain IL, Goldberg HA (2013) Graphite fibers and filaments. Springer Science & Business Media
43. Timoshenko S, Gere JM (2012) Theory of elastic stability. Dover Publications
44. Gong L, Young RJ, Kinloch IA, Haigh SJ, Warner JH, Hinks JA et al (2013) Reversible loss of Bernal stacking during the deformation of few-layer graphene in nanocomposites. ACS Nano 7(8):7287–7294
45. Frank O, Bousa M, Riaz I, Jalil R, Novoselov KS, Tsoukleri G et al (2012) Phonon and structural changes in deformed Bernal stacked bilayer graphene. Nano Lett 12(2):687–693

46. Mohiuddin T, Lombardo A, Nair R, Bonetti A, Savini G, Jalil R et al (2009) Uniaxial strain in graphene by Raman spectroscopy: G peak splitting, Grüneisen parameters, and sample orientation. Phys Rev B 79(20):205433
47. Androulidakis C, Koukaras EN, Frank O, Tsoukleri G, Sfyris D, Parthenios J et al (2014) Failure processes in embedded monolayer graphene under axial compression. Sci Rep 4:5271
48. Mei H, Huang R, Chung JY, Stafford CM, Yu H-H (2007) Buckling modes of elastic thin films on elastic substrates. Appl Phys Lett 90(15):151902–151903
49. Mei H, Landis CM, Huang R (2011) Concomitant wrinkling and buckle-delamination of elastic thin films on compliant substrates. Mech Mater 43(11):627–642
50. Meyer JC, Geim AK, Katsnelson MI, Novoselov KS, Booth TJ, Roth S (2007) The structure of suspended graphene sheets. Nature 446(7131):60–63
51. Fasolino A, Los J, Katsnelson MI (2007) Intrinsic ripples in graphene. Nat Mater 6(11):858–861
52. Lui CH, Liu L, Mak KF, Flynn GW, Heinz TF (2009) Ultraflat graphene. Nature 462(7271):339–341
53. Stolyarova E, Rim KT, Ryu S, Maultzsch J, Kim P, Brus LE et al (2007) High-resolution scanning tunneling microscopy imaging of mesoscopic graphene sheets on an insulating surface. Proc Natl Acad Sci 104(22):9209–9212

Chapter 4
Interfacial Friction and Adhesion Between Graphene and Silicon

4.1 Introduction

As mechanically robust, electrically conducting, atomically thin membrane, graphene promises to be an excellent candidate for next generation nanoelectromechanical systems (NEMS), ranging from nanoscale resonators, oscillators, switches, and valves that are applied in information processing, molecular manipulation, and sensing [1–6]. Accompanied with the miniaturization of electromechanical devices, the influence of van der Waals (vdW) forces that are practically insignificant at the macroscale tends to increase, thereby enhancing the attraction within nanomechanical structures, and potentially playing a dominant role in the device performance [7–9]. Critical issues hence appear concerning how to quantify these weak interfacial forces and what the impact of the vdW interface is on the nanoscale deformation of graphene-based devices. On the other hand, graphene being extremely thin with a large specific surface area is highly susceptible to interfacial forces and offers as an ideal candidate to study and characterize these forces [10, 11].

While engineering graphene-based NEMS is a recent area of study, the mechanical characterization of vdW-dominated adhesion properties of graphene has been explored for decades [12, 13]. Among diverse experimental approaches, bulging test (or blister test) is one of the most commonly used method. Zong et al. developed a particle intercalation method to characterize graphene adhesion. Gold or silver nanoparticles trapped at graphene/silicon interface act as point wedges to support circular blisters, allowing the SEM measurements of blister heights and radii. The average adhesion energy of the typical graphene/silicon interface in the high vacuum is found to be is measured to be 151 ± 28 mJ/m^2 [14]. Later, Koenig et al. extended the pressurized blister test to mechanically exfoliated graphene which were suspended over microcavities patterned in silicon oxide substrates. The pressure load can be precisely controlled and enables the delamination of graphene membrane. Based the atomic force microscopy (AFM) measurement of bubble profiles, the adhesion energy of graphene/SiO_2 interface was estimated to be 0.45 J/m^2 for a monolayer graphene and 0.31 J/m^2 for few-layers (2–5 layers). The reason for such a thickness

G. Wang, *Characterization and Modification of Graphene-Based Interfacial Mechanical Behavior*, Springer Theses, https://doi.org/10.1007/978-981-15-8029-1_4

dependency was deemed to be bending rigidity induced varying levels of conformation of the graphene membrane to the SiO_2 substrate [15]. Cao et al. reported an adhesion energy of 0.51 J/m^2 for CVD graphene transferred to copper film based on the blister test, where the blister height was determined by optical interference. A modified plate theory was proposed to account for residual stress effects and the rotational stiffness of the delamination front when relatively weak interactions are at play [16]. Based on the same experimental setup, Xin et al. further studied the adhesion properties of as-grown graphene on copper foil and the chemical treatment was performed to tailor the surface roughness. Counterintuitively, the rougher surface leads to higher adhesion energy, which was found to increase to 1.53 ± 0.11 J/m^2 [17]. However, the extensively used theoretical model typically assumes a clamped boundary at the interface around the bubble; however, the atomically smooth nature of graphene challenges this assumption.

In reality, slip events at vdW interfaces have been observed through both experiments and simulations. For the bulging test of monolayer graphene, Koenig et al. suggested possible sliding between graphene and SiO_2 substrate, which might be the origin of the scattered data points that deviate from the theoretical prediction at relatively high pressure loads [15]. Furthermore, the direct observation of wrinkling and buckling at the periphery of graphene bubbles offered additional evidence of the interfacial sliding [18]. By numerical simulations based on an atomistically informed continuum theory, Zhang et al. interpreted the wrinkle-to-buckle transition and proposed a strategy to control the surface morphology of buckle networks through tunable adhesive and frictional properties [19, 20]. Kitt et al. for the first time employed Raman spectroscopy to characterize such sliding behavior and extracted the sliding friction at the graphene/SiO_2 interface [21]. However, the impact of the interfacial slip on the mechanics of graphene membrane is still not well-understood, which is of particular importance to graphene-based NEMS involving the complex interplay between adhesive and frictional forces, thus calling for further studies.

In this chapter, we demonstrate a method for determining the interfacial shear stress via a pressurized bulging device by combining Raman and AFM measurements with a membrane analysis. We further examine such interfacial shear effect on the elastic stiffness and adhesion properties of graphene membrane. The mechanical mechanism underlying the coexistence of bubble and buckle was analyzed in detail, with an emphasis on the friction behavior at interface, which provides a framework to control complex out-of-plane motifs in supported graphene through the interfacial engineering [22].

4.2 Experimental Methods

4.2.1 *Preparation and Characterization of Suspended Graphene*

An array of circles with diameters of 3 and 5 μm was defined by photolithography on an oxidized silicon wafer with a silicon oxide thickness of 300 nm. Reactive ion etching was then used to etch the circles into cylindrical microcavities with a depth of around 300 nm. The distance between microcavities is 15 μm. Figure 4.1a, b show the characterization results by scanning electron microscopy (SEM) and atomic force microscopy (AFM) respectively. Graphene membranes were fabricated by the micromechanical cleavage method with a Scotch tape and then transferred onto the microcavities [23]. The morphology and lateral dimension were characterized by optical microscope and SEM, as shown in Fig. 4.1c, d. Here we focus on monolayer graphene, with the thickness identified by AFM and Raman spectroscopy (Fig. 4.1e, f). The AFM (Dimension icon, Veeco) measurement was performed in tapping mode with a scanning frequency of 1 Hz. The RTESP silicon tip was used the experiment, having a resonance frequency of 300 kHz and a cantilever spring constant of 40 N/m. The Raman spectrum was recorded using the 532 nm line of a diode-pumped solid-state laser. The laser power was kept below 0.5 mW on the sample to avoid laser-induced local heating of the sample. The focused laser spot was approximately 1 μm in diameter, and the spectral resolution was ~1 cm^{-1}.

4.2.2 *Pressurized Bulging Test*

The samples were placed into a pressure chamber to establish a pressure difference across the graphene membrane and the accuracy of the gauge was ~0.01 MPa. Nitrogen gas was pumped into the chamber to increase the external pressure to a specific level (p_0), as shown in Fig. 4.2a. Initially, the pressure inside the graphene-sealed microcavity (p_i) equals to the atmosphere pressure (p_{ext}~0.1 MPa) so that graphene membrane exhibits a flat morphology, as illustrated in Fig. 4.2b. As the pressure in chamber (p_0) increases, graphene membrane is first pushed downwards and then gradually recovers, mainly due to the gas diffusion through the SiO_2 substrate or graphene/SiO_2 interface into the microcavity. Commonly, it takes around one week to reach the pressure equilibrium, i.e. $p_i = p_0$. Once removed from the chamber, the internal pressure (p_i) would be higher than the atmospheric pressure (p_{ext}) outside so that the graphene membrane would bulge upwards and the gas volume would increase (Fig. 4.2c). If the pressure difference is high enough to overcome the interfacial adhesion energy, graphene membrane will delaminate from the SiO_2 substrate, as shown in Fig. 4.2d. Herein, the ideal gas law was used along with the assumption of isothermal expansion of the trapped gas with a constant number of molecules (N) in each bubble. As a result, we can obtain the internal pressure of the bubble by

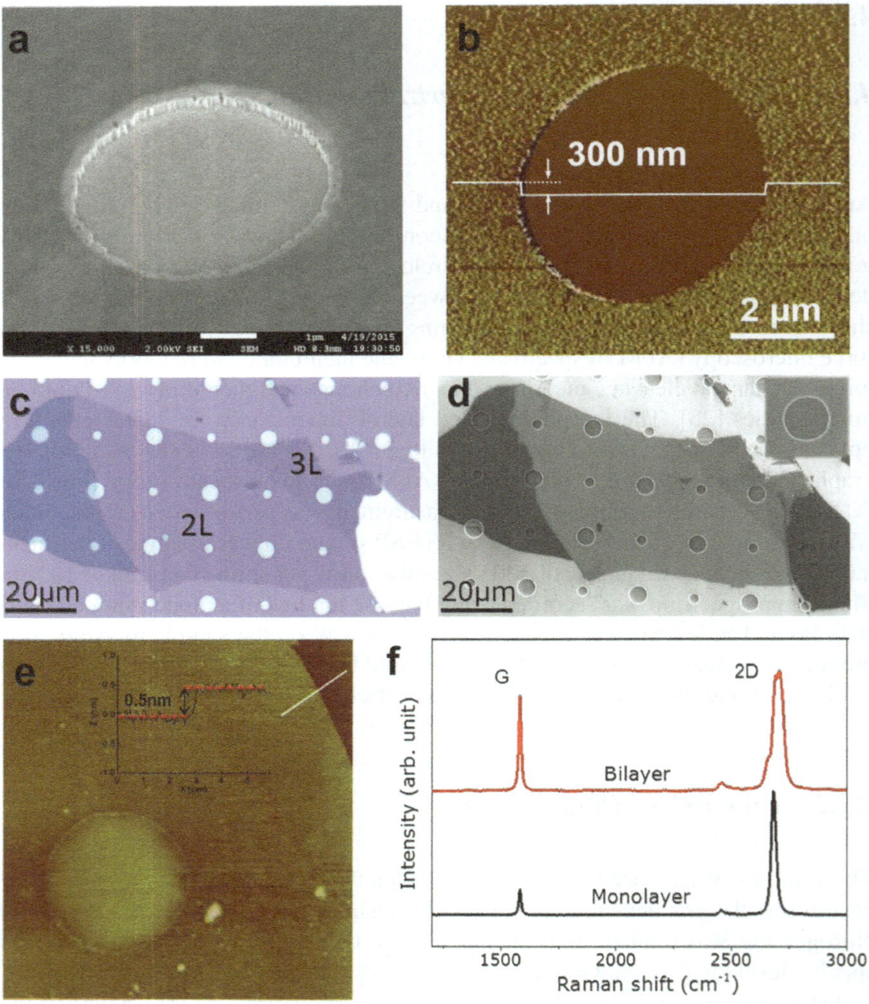

Fig. 4.1 **a** SEM and **b** AFM images of the patterned circular holes on SiO_2 substrate. **c** Optical and **d** SEM images of exfoliated graphene sheets on the prepatterned SiO_2 substrate. **e** Thickness of monolayer graphene characterized by AFM is ~0.5 nm. **f** Raman spectra of monolayer and bilayer graphene

$$p_0 V_0 = p_i (V_0 + V_b) \tag{4.1}$$

where V_0 is the initial volume of the hole and V_b is the volume of the bubble. Note that, the final pressure difference ($\Delta p = p_i - p_e$) was tuned through the applied pressure p_0 at an increment of 0.05 MPa, in case that graphene membrane collapses to the bottom of cavity. The bubble profile could be easily measured through AFM imaging

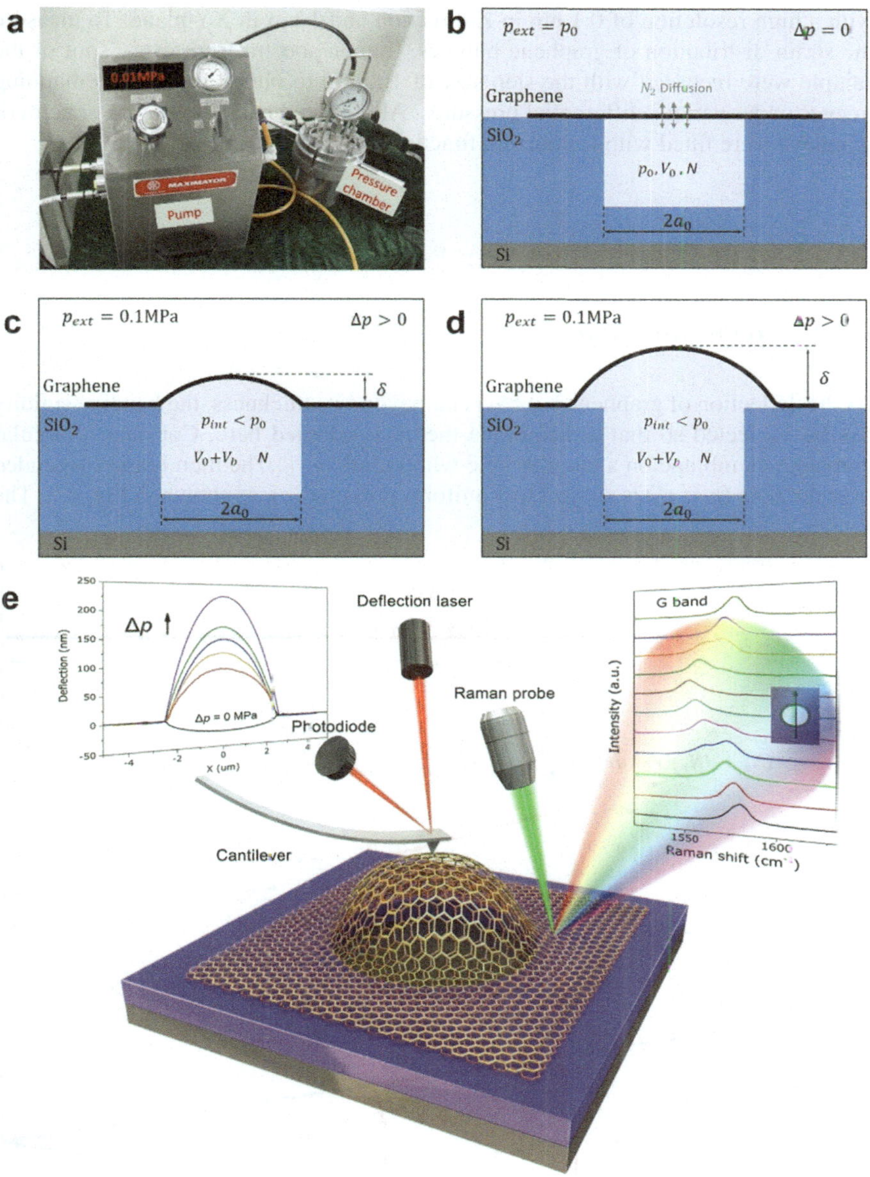

Fig. 4.2 **a** Pressure pump equipped with a pressure chamber. **b–d** The schematic of the bubbling process. **e** Schematic diagram of a graphene bulging device and characterizations. The left inset shows profiles of a graphene bubble with increasing pressures. The right inset shows Raman spectra of graphene G-band stacked vertically in the direction of a line-scan

with a high resolution of 0.1 nm in Z-direction and 1 nm in XY-plane. To measure the strain distribution of graphene bubbles, Raman spectra from every spot of the sample were recorded with the step size of 100 nm to obtain the Raman mapping images under various differential pressures. All featured bands in Raman spectra of graphene were fitted with Lorentzian functions to obtain the peak positions.

4.3 Bubble Deformation

4.3.1 Membrane Analysis

As the deflection of graphene bubble is far beyond its thickness, the bending rigidity can be neglected so that a membrane theory is adopted here. Consider a circular graphene membrane on a circular hole with radius of r_0. The membrane suspended over the hole ($r < r_0$) is subject to a uniform pressure (q), as shown in Fig. 4.3. The

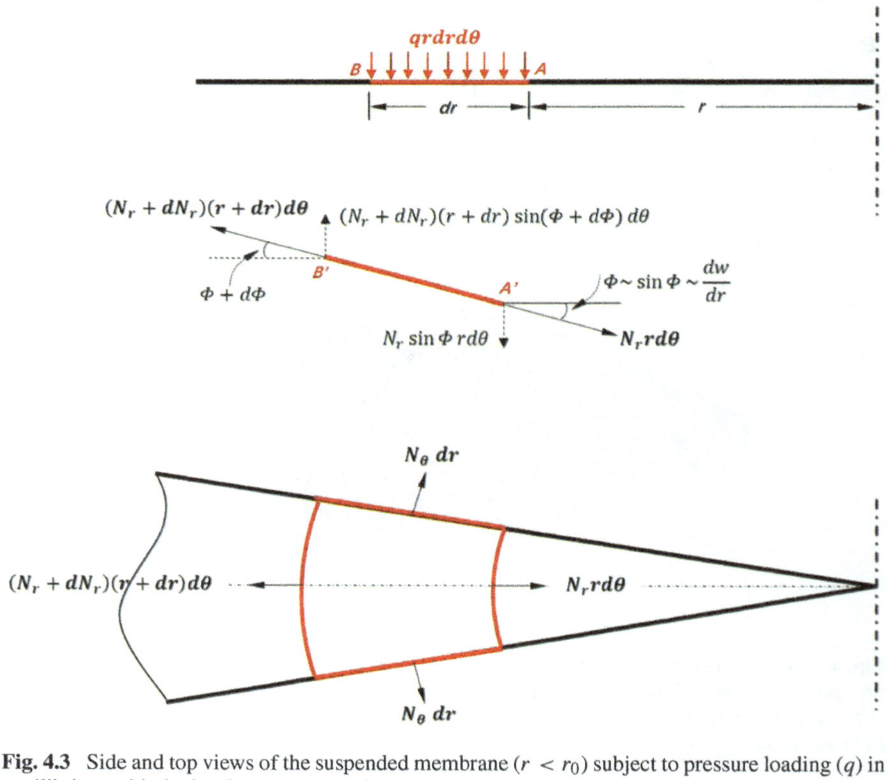

Fig. 4.3 Side and top views of the suspended membrane ($r < r_0$) subject to pressure loading (q) in equilibrium with the in-plane stress resultants

governing equations for the suspended membrane over the hole are [24, 25]:

$$\frac{d}{dr}(rN_r) - N_\theta = 0 \tag{4.2}$$

$$\frac{d}{dr}\left(rN_r\frac{dw}{dr}\right) + qr = 0 \tag{4.3}$$

where w is the lateral deflection, t is the membrane thickness, and q is the lateral loading intensity (i.e. the pressure difference across the graphene membrane Δp); $N_r = \sigma_r t$ and $N_\theta = \sigma_\theta t$ are, respectively, radial and circumferential stress resultants, where σ_r and σ_θ are radial and circumferential stresses.

For an elastic membrane, we have the stress–strain relationship [26]:

$$E\varepsilon_r = \sigma_r - \nu\sigma_\theta \tag{4.4}$$

$$E\varepsilon_\theta = \sigma_\theta - \nu\sigma_r \tag{4.5}$$

and strain–displacement relationship:

$$\varepsilon_r = \frac{du}{dr} + \frac{1}{2}\left(\frac{dw}{dr}\right)^2 \tag{4.6}$$

$$\varepsilon_\theta = \frac{u}{r} \tag{4.7}$$

where u is the radial displacement, E is Young's modulus, ν is Poisson' s ratio, ε_r and ε_θ are radial and circumferential strains, respectively. Combining Eqs. (4.2)–(4.7) leads to a nonlinear equation:

$$N_r^2\frac{d}{dr}\left[\frac{1}{r}\frac{d}{dr}(r^2N_r)\right] + \frac{Etq^2}{8}r = 0 \tag{4.8}$$

Following Hencky's approach [27], the solutions to $N_r(r)$, $N_\theta(r)$ and $w(r)$ are assumed to take the form:

$$N_r = \left(\frac{Etq^2{r_0}^2}{64}\right)^{\frac{1}{3}}\sum_{n=0}^{\infty} b_{2n}\left(\frac{r}{r_0}\right)^{2n} \tag{4.9}$$

$$N_\theta = \left(\frac{Etq^2{r_0}^2}{64}\right)^{\frac{1}{3}}\sum_{n=0}^{\infty}(2n+1)b_{2n}\left(\frac{r}{r_0}\right)^{2n} \tag{4.10}$$

$$w = \left(\frac{q{r_0}^4}{Et}\right)^{\frac{1}{3}}\sum_{n=0}^{\infty} a_{2n}\left[1-\left(\frac{r}{r_0}\right)^{2n}\right] \tag{4.11}$$

where seven terms ($n = 0$–6) are included in the polynomial expansions. Substituting Eq. (4.9) into Eq. (4.8), all the coefficients b_{2n} can be determined in terms of one parameter, b_0, corresponding to the stress coefficient at the center of the hole [16, 21]. That is,

$$b_2 = -\frac{1}{b_0{}^2};\quad b_4 = -\frac{2}{3b_0{}^5};\quad b_6 = -\frac{13}{18b_0{}^8};\quad b_8 = -\frac{17}{18b_0{}^{11}};$$
$$b_{10} = -\frac{37}{27b_0{}^{14}};\quad b_{12} = -\frac{1205}{567b_0{}^{17}} \tag{4.12}$$

Similarly, substituting Eq. (4.11) into Eq. (4.3) yields.

$$a_0 = \frac{1}{b_0};\quad a_2 = \frac{1}{2b_0{}^4};\quad a_4 = \frac{5}{9b_0{}^7};\quad a_6 = \frac{55}{72b_0{}^{10}};\quad a_8 = \frac{7}{6b_0{}^{13}};$$
$$a_{10} = \frac{205}{108b_0{}^{16}};\quad a_{12} = \frac{17051}{5292b_0{}^{19}} \tag{4.13}$$

The coefficient b_0 can be determined numerically for specific Poisson's ratios and boundary conditions, for instance, for an elastic membrane clamped at the edge and $\nu = 0.165$, $b_0 = 1.66$. In general, however, b_0 depends on the interfacial shear stress in the supported region and the applied pressure difference in the suspended region.

According to Eq. (4.11), two measurable quantities (i.e. pressure difference, Δp across the membrane and the maximum deflection at the center of the hole, $\delta = w|_{r=0}$) follow a simple relation:

$$\Delta p = kEt\frac{\delta^3}{r_0{}^4} \tag{4.14}$$

where $k = \left(\frac{1}{b_0} + \frac{1}{2b_0{}^4} + \frac{5}{9b_0{}^7} + \cdots\right)^{-3}$ is a dimensionless coefficient that depends on b_0. As a reference, we define $k = k_0$ for the clamped edge condition, which equals 3.10 for $\nu = 0.165$. In general, as b_0 depends on the interfacial shear stress and the applied pressure difference, so does the coefficient k.

4.3.2 Combined Raman Spectroscopy-AFM Characterization

AFM in the standard tapping mode was utilized to measure the shape of the graphene bubbles, including the maximum deflection (δ) and the radius (r_0). The three-dimensional rendering as well as the cross section of the AFM images is presented as shown in Fig. 4.4, showing the evolution of the deflection profiles of a graphene bubble with increasing pressure. Initially, the graphene membrane appears to sag in the hole due to the van der Waals attraction to the inner walls (Fig. 4.4a) [1, 28]. Increasing the pressure causes the graphene bubble to form and grow in height,

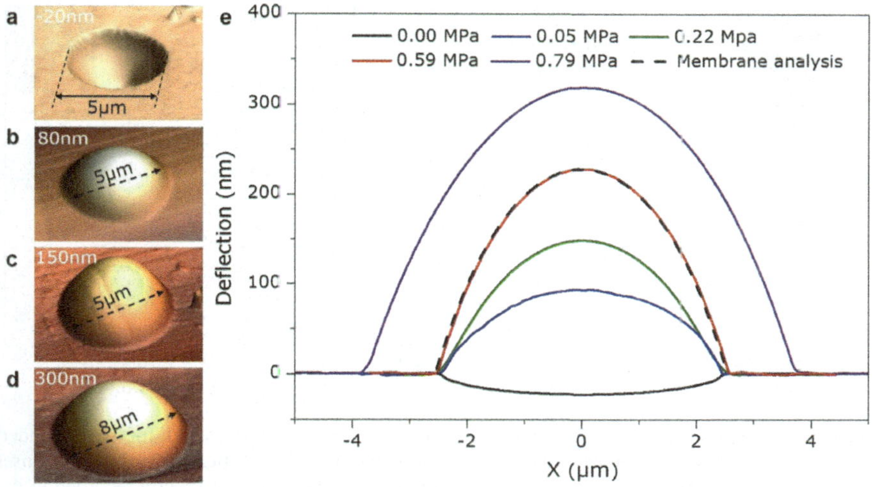

Fig. 4.4 **a–d** Three-dimensional AFM images show the evolution of the graphene bubble with increasing deflection and eventually delamination. **e** AFM profiles of the graphene bubble at different pressure differences (Δp). The deflection was measured by AFM along a line passing through the center of the hole. The dashed line was obtained from our membrane analysis as detailed later

until the delamination from the SiO_2 substrate, leading to an increase in bubble radius (Fig. 4.4b–d). The bubble profile as depicted in Fig. 4.4e exhibits a sphere-cap shape, recalling the membrane deformation characteristics [25, 29, 30]. In contrast to the common constant pressure bulging test, which features unstable crack growth at the onset of delamination, herein, the number of molecules in the microcavity is constant due to the high impermeability of graphene [31, 32]. This hence allows the reliable measurements of stable bubble deformation.

Figure 4.5a presents the Raman spectrum of monolayer graphene inside and outside the microcavity under different pressures. At $\Delta p = 0$ MPa, Raman spectrum of the graphene at the suspended region exhibits a subtle difference from that of the supported region. Specifically, Raman G-band of the suspended region is appreciably downshifted by ~2.6 cm^{-1}, compared with that measured on the supported region (1580.6 vs. 1583.2 cm^{-1}). The slight downshift of Raman G-band is attributed to the slight tension resulted from an initial sag of the suspended part as presented in Fig. 4.4a. Once the applied pressure was increased up to $\Delta p = 0.14$ MPa, the Raman G-band frequency shifted significantly from 1579 to 1560 cm^{-1} for the suspended graphene, largely due to the strain effects induced by formation of the bubble, accompanied with a more significant downshifts of the 2D-band from 2670 to 2632 cm^{-1}. Even larger downshifts of the Raman peaks are visible at higher pressures. When the pressure difference rises to 0.35 MPa, the redshift of Raman G-band reaches as high as 55 cm^{-1}, as shown in Fig. 4.5b. Meanwhile, the band width obtained from the suspended region (13 cm^{-1}) is greater than that of supported region (7 cm^{-1}), which may be explained by the doping effect of the supported graphene as it has

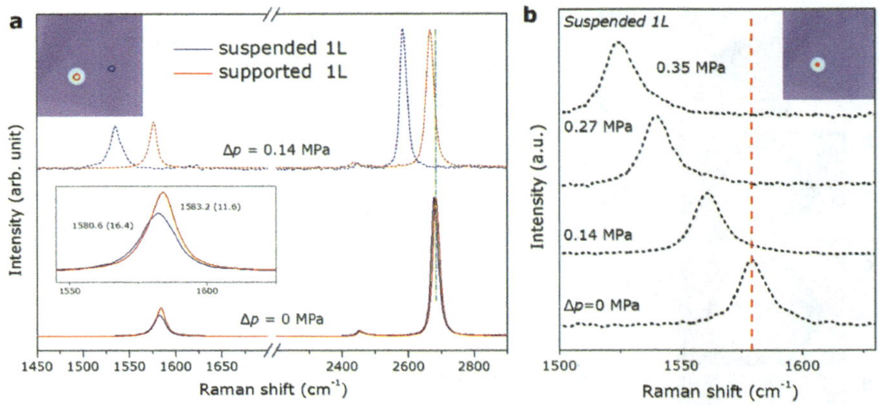

Fig. 4.5 **a** Raman spectra recorded on the monolayer (1L) bulging devices both for the suspended (blue line) and supported regions (red line) at different pressures. The optical image in the up-inset shows the locations at which Raman measurements were taken. **b** Raman spectra recorded in the center of suspended region on the monolayer graphene at different pressures

been demonstrated that the hole doping induced by the substrate tends to cause the sharpening of Raman G-band [33, 34]. Note that, the initial strain and doping effects become negligible after applying a pressure difference, and thus the pressurized bubbling was the primary mechanism for the strains as well as the corresponding evolution of the Raman features in the suspended and supported regions.

4.4 Interfacial Shear Between Graphene and Silicon

4.4.1 Propagation of Shear Zone Surrounding the Bubble

It is noticed that, the bubble deformation induces not only substantial downshifts in the suspended region, but also slight downshifts in the supported region (Fig. 4.5a). As illustrated Fig. 4.6a, the applied pressure inside the hole pushes the graphene upwards and induces a tension at the edge of the hole, pulling the supported graphene towards the hole. To resist the relative sliding at the interface, an interfacial shear stress, τ_1, is generated between graphene and the substrate. Leveraging high strain sensitivity of Raman peak of graphene [35–37], we then employ Raman maps plotting the frequency shifts of G-band to experimentally identify the interfacial (graphene/SiO_2) shear behaviors outside the microcavity. Figure 4.6b shows the initial Raman G-band frequency mappings of as-prepared monolayer graphene samples at $p_0 = p_e$, exhibiting uniform distributions outside the hole whereas small frequency shifts are observable at the center of the hole. This indicates a small strain inside the hole and nearly zero strain outside. Subsequently, as the monolayer graphene was bulged

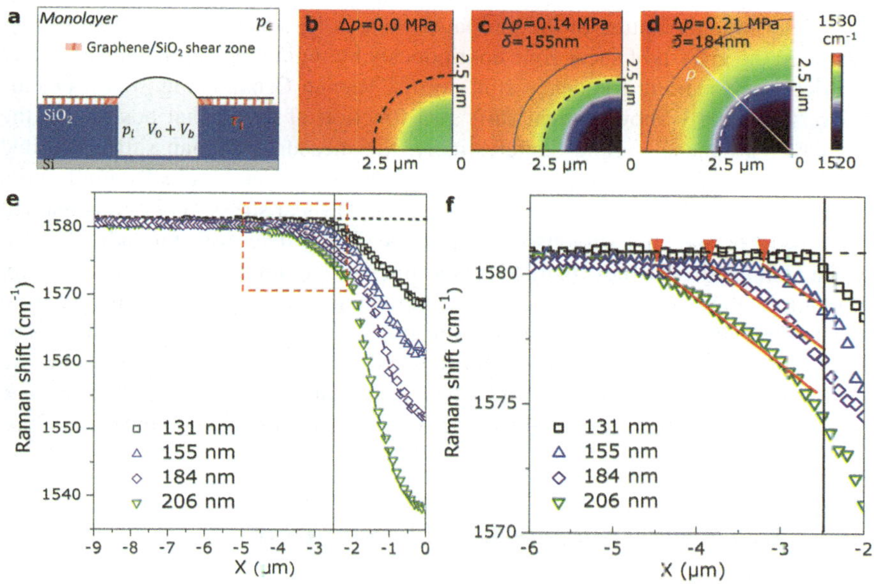

Fig. 4.6 **a** Schematic diagrams of bulged monolayer graphene samples. **b–d** Raman contour maps of G-band frequency in one quadrant revealed the strain distributions in monolayer samples in both suspended and supported regions at different pressures. The cross sections of Raman contour maps are shown in **e**. **f** The variation of the G-band frequency near the edges of the hole. Solid vertical lines are positioned at the edges of the hole, and the dashed horizontal lines indicate the frequency of the G-band at zero strain

upwards with maximum deflection $\delta = 155$ nm at the center under $\Delta p = 0.14$ MPa, the G-band red shifts significantly inside the hole (Fig. 4.6c). Meanwhile, the G-band of the supported graphene also exhibited red shifting, exhibiting a strain gradient (denoted as the color gradient) in an annular region outside the hole. That is a clear indication of interfacial shear deformation between supported graphene and the SiO_2 substrate outside the hole. To quantitatively describe the propagation of the interfacial shear zone with a finite outer radius, We define a dimensionless parameter $\rho = R/r_0$ as the ratio between the outer radius of the shear zone (R, the solid line) and the radius of the hole (r_0, the dashed line). Within the shear zone ($r_0 < r < R$), Raman redshifts or the corresponding strain level decreases with the distance from the edge of the hole until the G-band returns to the unstrained level. At $\Delta p = 0.21$ MPa, the Raman mapping (Fig. 4.6d) shows larger redshifts in both suspended and supported regions. Moreover, the shear zone in the supported region grows controllably outwards with a larger ρ, implying that more interfacial area outside the hole participated in the shear deformation to balance the increased pressure inside the hole. Importantly, bulging test is considered as a conventional approach to measure the interfacial adhesion energy between the film and substrate [15–17, 38], but we found that the critical pressure for interfacial slippage is much lower than that for delamination.

To determine the nature of the interlayer shear zone outside the hole, we further controlled the growth of shear zone continuously by tuning the applied pressure and recorded line-scanning frequency shifts of the Raman G-band (Fig. 4.6e). Details in Fig. 4.6f clearly show Raman shift as a function of the spatial position in the monolayer graphene samples. Interestingly, the measured Raman shift exhibited nearly linear characteristics near the edge of the shear zone, typically implying a constant strain gradient due to a constant interfacial shear stress in one-dimensional scenarios [39]. Such linearity also enabled us to clearly determine the radius ratio ρ for the annular shear zones (marked by red lines and symbols), with respect to Δp. In principle, the biaxial strain and stress in graphene may also be determined from the measured Raman shifts for monolayer graphene samples, which could then be used to determine the shear stresses for the graphene/SiO_2 (τ_1) interface. Here, we determine the interfacial shear stress based on the measurement of the shear zone size (ρ) from Raman mapping, which does not require precise strain measurements.

4.4.2 *Measurement of Interfacial Shear Stress*

Previous studies assuming clamped boundary conditions represent the limiting case with an infinite shear stress at the interface between the membrane and its substrate so that the deformation of the membrane is negligible outside the hole [15, 18, 40, 41]. Such an assumption is reasonable when the interfacial shear stresses (frictional force per unit area), τ, is large enough to resist interfacial sliding so that the membrane could be considered to be firmly anchored to the substrate before delamination occurs. However, in contrast to the relatively strong adhesion of graphene to its substrate, the interfacial shear stress between graphene and its substrate is expected to be low so that sliding could be readily actuated. In this case, outside the hole the supported graphene membrane is constrained to in-plane deformation with radial symmetry and a constant interfacial shear stress (frictional resistance) [21, 42]. We herein solve the plane stress problem outside the hole with a finite interfacial shear stress, allowing further examination of the experimental data and simulation results.

Assuming no buckling or any out-of-plane deflection (i.e., $w = 0$) for the supported membrane outside the hole, the governing equations include an in-plane equilibrium equation (Fig. 4.7) [21, 43]:

$$\frac{d}{dr}(rN_r) - N_\theta + \tau r = 0 \tag{4.15}$$

and the strain–displacement relationship:

$$\varepsilon_r = \frac{du}{dr} \tag{4.16}$$

$$\varepsilon_\theta = \frac{u}{r} \tag{4.17}$$

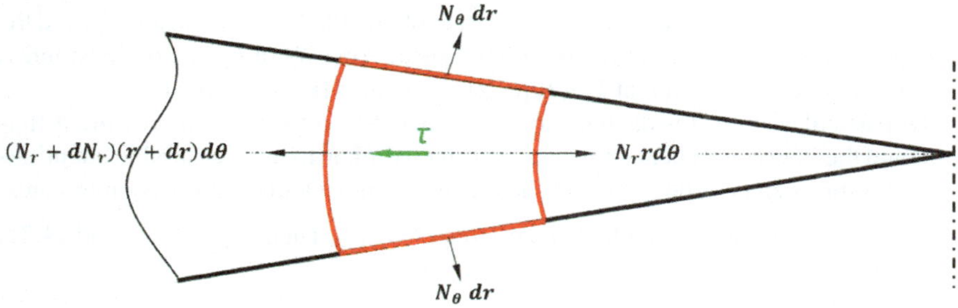

Fig. 4.7 Top view of the supported membrane ($r > r_0$) showing the in-plane stress resultants in equilibrium with an interfacial shear stress

as well as the elastic stress–strain relationship in Eqs. (4.4) and (4.5). Combination of Eqs. (4.4)–(4.5) and Eqs. (4.15)–(4.17) leads to a linear stress equation:

$$\frac{d}{dr}\left[N_r + \frac{d}{dr}(rN_r)\right] = -(2+\nu)\tau \tag{4.18}$$

With the boundary condition that the in-plane stresses decay to zero ($N_r = N_\theta = 0$) at $r = \rho r_0$, Eq. (4.18) can be solved exactly to obtain:

$$N_r = \tau r_0\left[(2+\nu)\left(-\frac{1}{3}\frac{r}{r_0} - \frac{\nu-1}{6(2+\nu)}\rho^3\frac{r_0^2}{r^2}\right) + \frac{1+\nu}{2}\rho\right] \tag{4.19}$$

$$N_\theta = \tau r_0\left[(2+\nu)\left(-\frac{1+2\nu}{3(2+\nu)}\frac{r}{r_0} + \frac{\nu-1}{6(2+\nu)}\rho^3\frac{r_0^2}{r^2}\right) + \frac{1+\nu}{2}\rho\right] \tag{4.20}$$

At the edge of the hole ($r = r_0$), the radial stress and displacement are continuous. As a result, both the strain and stress components are continuous, i.e., $N_r^{in} = N_r^{out}$ and $N_\theta^{in} = N_\theta^{out}$ at $r = r_0$. Based on Eqs. (4.9)–(4.10) for the suspended membrane over the hole and Eqs. (4.19)–(4.20) outside the hole, the continuity conditions at the edge of the hole lead to:

$$\tau r_0\left[(2+\nu)\left(-\frac{1}{3} - \frac{\nu-1}{6(2+\nu)}\rho^3\right) + \frac{1+\nu}{2}\rho\right] = \left(\frac{Etq^2 r_0^2}{64}\right)^{\frac{1}{3}}\left(b_0 - \frac{1}{b_0^2} - \frac{2}{3b_0^5} - \cdots\right) \tag{4.21}$$

$$\tau r_0\left[(2+\nu)\left(-\frac{1+2\nu}{3(2+\nu)} + \frac{\nu-1}{6(2+\nu)}\rho^3\right) + \frac{1+\nu}{2}\rho\right] = \left(\frac{Etq^2 r_0^2}{64}\right)^{\frac{1}{3}}\left(b_0 - \frac{3}{b_0^2} - \frac{10}{3b_0^5} - \cdots\right) \tag{4.22}$$

Equations (4.21) and (4.22) can be solved to obtain the Hencky constant b_0 and the ratio ρ for the outer radius of the interfacial shear zone. With b_0, the dimensionless pressure-deflection coefficient k in Eq. (4.14) can also be determined.

Of particular interest is the dependence of ρ and k on the applied pressure difference and the interfacial shear stress as they directly relate to the measurable quantities (i.e., pressure, center deflection, and shear zone radius). Define a dimensionless pressure $\bar{q}=\frac{qr_0}{Et}$ and a dimensionless shear stress $\bar{\tau}=\frac{\tau r_0}{Et}$. Then Eqs. (4.21) and (4.22) are converted to

$$\left[(2+\nu)\left(-\frac{1}{3}-\frac{\nu-1}{6(2+\nu)}\rho^3\right)+\frac{1+\nu}{2}\rho\right]=\frac{1}{4}\frac{\left(\bar{q}\right)^{\frac{2}{3}}}{\bar{\tau}}\left(b_0-\frac{1}{b_0{}^2}-\frac{2}{3b_0{}^5}\cdots\right) \tag{4.23}$$

$$\left[(2+\nu)\left(-\frac{1+2\nu}{3(2+\nu)}+\frac{\nu-1}{6(2+\nu)}\rho^3\right)+\frac{1+\nu}{2}\rho\right]$$
$$=\frac{1}{4}\frac{\left(\bar{q}\right)^{\frac{2}{3}}}{\bar{\tau}}\left(b_0-\frac{3}{b_0{}^2}-\frac{10}{3b_0{}^5}\cdots\right) \tag{4.24}$$

It is found that both b_0 and ρ depend on one combined parameter, $\zeta=\bar{q}^{\frac{2}{3}}/\bar{\tau}$. In addition, they depend on Poisson's ratio, which is assumed to be a constant (0.165) for graphene. We plot the results for b_0, ρ and k versus ζ in Fig. 4.8. Obviously, in the presence of interfacial shear deformation, k is no more a constant but changes with pressure and interfacial shear resistance. Evidently, when $\zeta\to 0$ (i.e. $\tau\to\infty$ or $q\to 0$), the results approach the limit for a clamped edge with $k=3.10$ and $\rho=1$. However, with a finite interfacial shear stress τ, ρ increases and k decreases as the applied pressure difference increases. As a result, it is possible to determine the interfacial shear stress based on the present membrane analysis and measurements for monolayer graphene bubbles, that is, the radius of interfacial shear zone (ρ) and the maximum deflection of graphene bubble (δ) versus applied pressure difference Δp. Furthermore, based on Eq. (4.14) and $k=\left(\frac{1}{b_0}+\frac{1}{2b_0{}^4}+\frac{5}{9b_0{}^7}+\cdots\right)^{-3}$, b_0 is actually a parameter associated with the stiffness of graphene. Increased pressure or decreased interfacial friction would cause more significant slippage of graphene on substrate, corresponding to the reduction in b_0 and k which means the softening of the Young's modulus (Fig. 4.8a, b). Likewise, a faster expanding of the shear zone can also be expected as shown in Fig. 4.8c.

Above-developed membrane analysis highlights the effect of interfacial friction on the propagation of shear zone as well as bubble deformation of graphene. Figure 4.9a shows that the interfacial shear zone grows with applied pressure difference for various interfacial shear stresses. Not surprisingly, a larger interfacial shear stress results in smaller shear zone growth. Specifically, when the shear stress is infinite

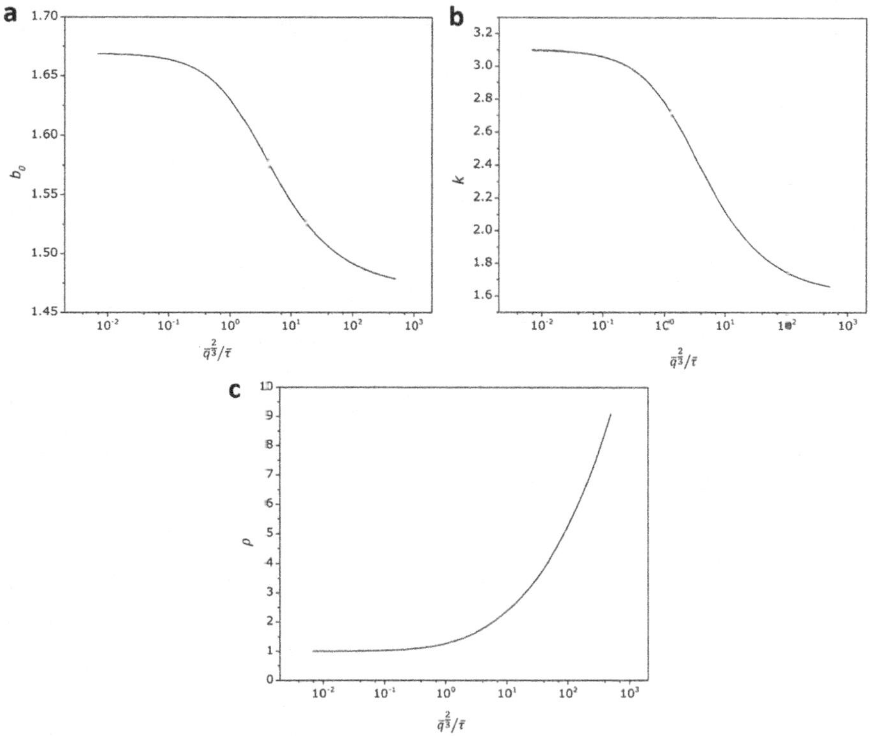

Fig. 4.8 **a** The dependency of Hencky constant b_0, **b** radius ratio ρ of the interfacial shear zone, and **c** pressure-deflection coefficient k on $\bar{q}^{-\frac{2}{3}} / \bar{\tau}$

(clamped boundary condition), the graphene/SiO_2 interface will be fixed upon the edge of the hole, with zero interfacial shear deformation and no shear zone (i.e. $\rho = 1$) regardless of the applied pressure difference. This leads to a constant k (defined as $k_0 = 3.10$ for graphene) and hence a linear relation between δ^3 and Δp, with a slope proportional to the in-plane stiffness of graphene Et. As the interfacial shear stress decreases to a finite value, the relationship between Δp and δ^3 becomes slightly nonlinear as shown in Fig. 4.9b. k would be lower than k_0 and the corresponding slope is larger than $1/Et$, while the difference between k and k_0 increases as the applied pressure increases. As a result, part of the deflection of the graphene bubble stems from interfacial sliding outside the hole especially at high applied pressure. Therefore, if interfacial sliding indeed occurred but were overlooked in the previous studies using clamped boundary conditions, the deflection of the graphene bubble would have been underestimated, or equivalently, the in-plane stiffness of graphene (Et) would have been underestimated by matching the deflection-pressure relation.

We could hence determine the interfacial shear stress based on the measurements of δ and ρ with respect to Δp in bulging tests of monolayer graphene. In Fig. 4.9c,

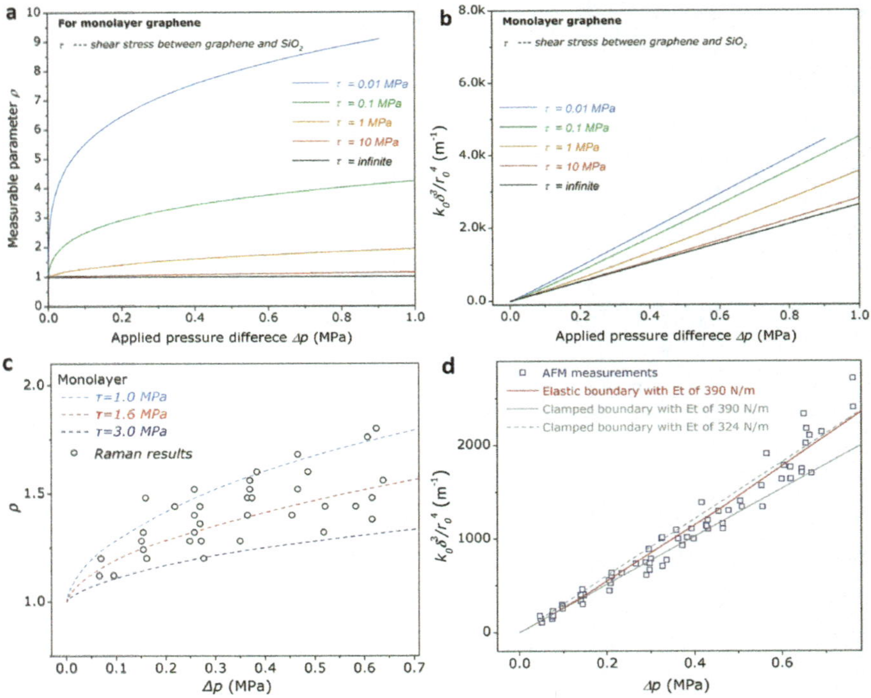

Fig. 4.9 The radius ratio ρ of the interfacial shear zone versus the applied pressure difference for different values of interfacial shear stress in **a** theoretical analysis and **c** experimental results. The Normalized center deflection $k_0\delta^3/r_0{}^4$ versus the applied pressure difference in **b** theoretical analysis and **d** experimental results

we plot ρ versus Δp for monolayer graphene bubbles. Following the theoretical analysis as discussed above, τ_1 is extracted with values in the range of 1–3 MPa, in good agreement with previous reports for the graphene/SiO_2 interfaces (0.25–2 MPa) [21, 44] and graphene/polymer interfaces (0.3–2.3 MPa) [45–47]. We note that the average in-plane stiffness Et obtained in Fig. 4.9c is slightly higher than previously reported values, which might be attributed to the effect of interfacial shear sliding outside the hole that was previously overlooked (Fig. 4.9d). In comparison with the prediction assuming no shear deformation outside the hole, the prediction with a finite shear stress ($\tau_1 = 1.64$ MPa) offers a better agreement with the AFM measurements, especially at relatively high pressure levels where more significant shear deformation occurred.

Based on the present membrane analysis, we propose a procedure to determine both the interfacial shear stress (τ_1) and the in-plane stiffness (Et) of monolayer graphene from the measurements of center deflection (δ) and the shear zone radius (ρ) in the supported region. First, the key parameter $\zeta_1 = \bar{q}^{-\frac{2}{3}}/\bar{\tau}_1$ can be determined by

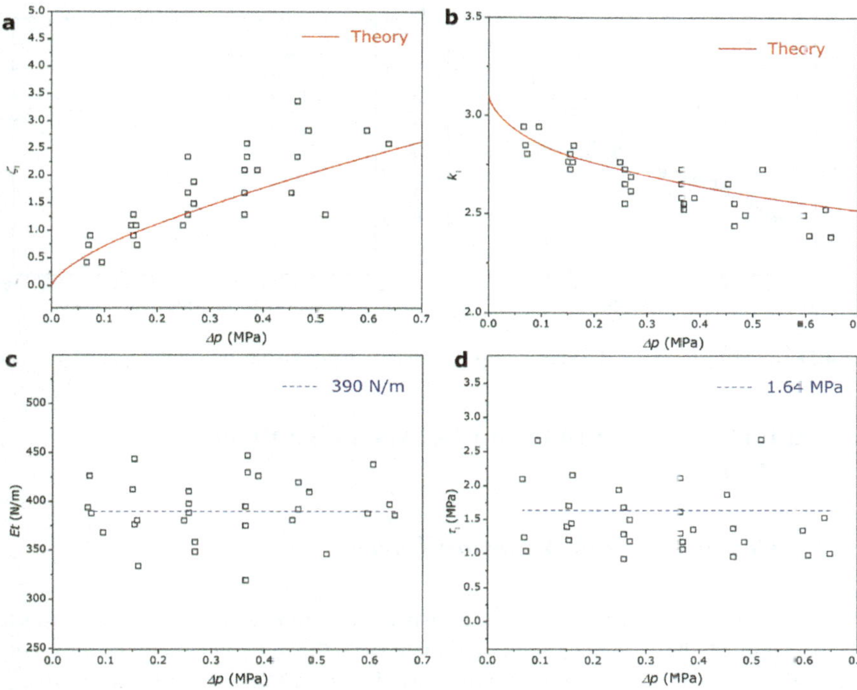

Fig. 4.10 The parameters extracted from monolayer measurements: **a** ζ_1 versus Δp, **b** k_1 versus Δp, **c** Et versus Δp, and **d** τ_1 versus Δp. The red lines in (**a**) and (**b**) are derived from the master curves in Fig. 4.8 using the average values of Et and τ_1 from (**c**) and (**d**), respectively

the master curve in Fig. 4.8b for each measured radius (ρ_1) of the shear zone at each applied pressure (Fig. 4.10a), where the subscript 1 denotes the interface between graphene and the SiO_2 substrate. Then, the pressure-deflection coefficient k_1 can be determined by Fig. 4.8c for the same parameter ζ_1 (Fig. 4.10b). Correspondingly, with measurements of the center deflection (δ) and the applied pressure ($q = \Delta p$), we can determine the in-plane stiffness Et based on Eq. (4.11), as shown in Fig. 4.10c. Finally, the interfacial shear stress can be determined as $\bar{\tau}_1 = q^{-\frac{2}{3}}/\zeta_1$ and $\tau_1 = \bar{\tau}_1 Et/r_0$, which is averaged as 1.64 MPa (Fig. 4.10d).

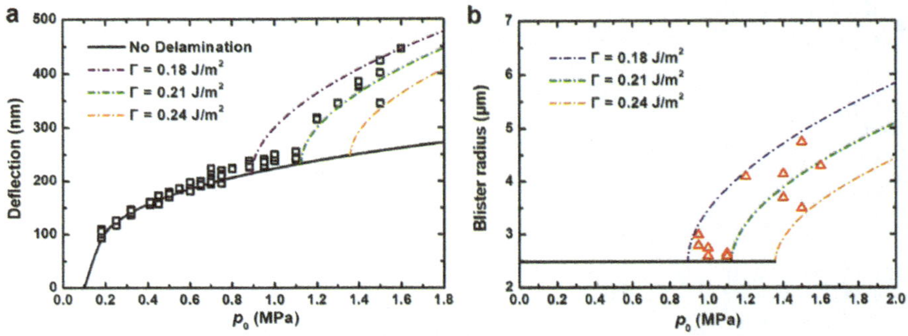

Fig. 4.11 **a** The deflection and **b** the radius of graphene bubble as a function of the applied pressure

4.5 Interfacial Delamination Between Graphene and Silicon

4.5.1 Measurement of Adhesion Energy

When the pressure difference across the suspended graphene membrane increases to the critical level, the graphene will delaminate from the outer perimeter of the cavity and stably expand outward. For instance, as shown in Fig. 4.11a, b, as interfacial delamination occurs at $p_0 = 1.2$ MPa, the bubble height displays an abrupt increase accompanied with a jump in bubble radius. Since the balance between adhesion (which favors large areas of contact) and stretching energy (which diminishes in blisters of large radius) dictates a constant aspect ratio of graphene, we are able to estimate the adhesion energy based on the AFM measurement of the bubble profile [37, 48].

In reality, the bulging test has long been a commonly used approach to measure the interfacial toughness or adhesion between coating film and the substrate [49, 50]. Compared to other traditional testing methods (such as peeling test and scratch test) [51–54], the advantage of the bulging test lies in the circumvention of excessive plastic deformation and structural damage. However, traditional bulging tests were loaded with constant pressure, typically initiating the delamination in an unstable manner once the critical pressure is reached [55]. Instead, for the delamination of graphene bubble with constant number of gas molecules, the pressure difference will on one hand push the bubble periphery to expand outward; on the other hand, the increasing bubble volume will reduce the internal pressure according to the ideal gas law, thereby mitigating the driving force to slow down the bubble growth. Such a feedback mechanism hence accounts for the stable delamination of graphene bubble which provides an excellent platform for the measurement of the adhesion energy between graphene and other materials. Recently, Boddeti et al. claimed that the stable delamination also requires $V_0 < 2V_b$ which guides the geometric design of the microcavity [30]. Specifically, too large cavity depths easily induce the uncontrollable delamination.

To determine the adhesion energy, here, we model the graphene bubble as a thermodynamic system to developing relations among the geometry, pressure loading and the interface adhesion energy. The free energy of the system is

$$U_T = U_E + U_G + U_P + U_A \tag{4.25}$$

U_T is the total free energy, U_E is the strain energy stored in the graphene bubble, U_G is the free energy change associated with expansion of gas molecules within the bubble, U_P is the free energy change of the external environment, U_A is the adhesion energy between graphene and substrate. Assuming the bubble growth process is quasistatic, we have the strain energy:

$$U_E = \iint N_i d\varepsilon_i dA = \Delta p V_b / 4 \tag{4.26}$$

where N_i is the membrane force resultant, ε_i is the associated strain, dA is an infinitesimal element of membrane cross sectional area, Δp is the pressure difference and V_b is the volume of bubble. Once the bubble is formed, the gas volume increases from V_0 to $V_0 + V_b$ and the pressure correspondingly decreases from p_0 to $p_i = \frac{p_0 V_0}{V_0 + V_b}$. Thus, the change of the free energy is

$$U_G = -\int \Delta p dV = -p_0 V_0 \ln\left(\frac{V_0 + V_b}{V_0}\right) \tag{4.27}$$

where p_0 is the applied pressure and V_0 is the volume of microcavity. As the blister expands by V_b, the volume of the surroundings decreases by an equal amount. Assuming the surroundings are maintained at an atmosphere pressure (p_e), the free energy then changes by

$$U_P = \int p_e dV = p_e V_b \tag{4.28}$$

With a given adhesion energy per unit area Γ, U_A is expressed as

$$U_A = \int \Gamma dA = \Gamma \pi (a^2 - r_0^2) \tag{4.29}$$

where a is the delaminated bubble radius. Combining Eqs. (4.25)–(4.29), we have the total free energy

$$U_T = \frac{\Delta p V_b}{4} - p_0 V_0 \ln\left(\frac{V_0 + V_b}{V_0}\right) + p_e V_b + \Gamma \pi (a^2 - r_0^2) \tag{4.30}$$

In particular, the volume of the bubble can be obtained by integral of the bubble profile based on Eq. (4.11):

$$V_b = \int w(r)2\pi r dr = \int_0^a \left(\frac{\Delta p a^4}{Et}\right)^{\frac{1}{3}} \sum_{n=0}^{\infty} a_{2n}\left[1 - \left(\frac{r}{a}\right)^{2n}\right] 2\pi r dr = C\pi a^2\delta \tag{4.31}$$

where C is a parameter as a function of a_0. The equilibrium configurations require the extrema of U_T to be zero. By substituting Eq. (4.31) into Eq. (4.30),

$$\frac{dU_T}{da} = \frac{V_b}{4}\frac{\mathrm{d}\Delta p}{da} - \frac{3\Delta p}{4}\frac{\mathrm{d}V_b}{da} + 2\Gamma\pi a = 0 \tag{4.32}$$

The combination of Eqs. (4.11) and (4.31) gives:

$$\frac{\mathrm{d}V_b}{da} = \left.\frac{\partial V_b}{\partial \Delta p}\right|_a \frac{\partial \Delta p}{\partial a} + \left.\frac{\partial V_b}{\partial a}\right|_{\Delta p} = \frac{V_b}{3\Delta p}\frac{\partial \Delta p}{\partial a} + \left.\frac{\partial V_b}{\partial a}\right|_{\Delta p} \tag{4.33}$$

substituting Eq. (4.33) into Eq. (4.32), we have

$$\frac{dU_T}{da} = -\frac{3\Delta p}{4}\left.\frac{\partial V_b}{\partial a}\right|_{\Delta p} + 2\Gamma\pi a = 0 \tag{4.34}$$

Reusing the ideal gas equation, we finally obtain

$$\Gamma = \frac{5C}{4}\Delta p\delta = \frac{5C}{4}\left(\frac{p_0 V_0}{V_0 + V_b} - p_e\right)\delta \tag{4.35}$$

Obviously, only if the bubble deflection and radius are measured, the adhesion energy can be determined. Figure 4.12a summarize the adhesion energy based on the AFM measurements of more than 40 graphene bubbles, giving an average of 0.19 ± 0.02 J/m^2 that is consistent with previous repost ranging from 0.14 to 0.45 J/m^2 [15, 18, 30]. More importantly, this value is comparable to the adhesion energy between liquid/solid [56] while much higher than that between solid/solid [7, 57–60], thus highlighting a liquid-like conformation of graphene over the SiO_2.

4.5.2 *Interplay Between Interfacial Slippage and Delamination*

Following the discussion in Sect. 4.4, interfacial shear deformation is demonstrated to influence the elastic response of graphene bubble. A question naturally arises

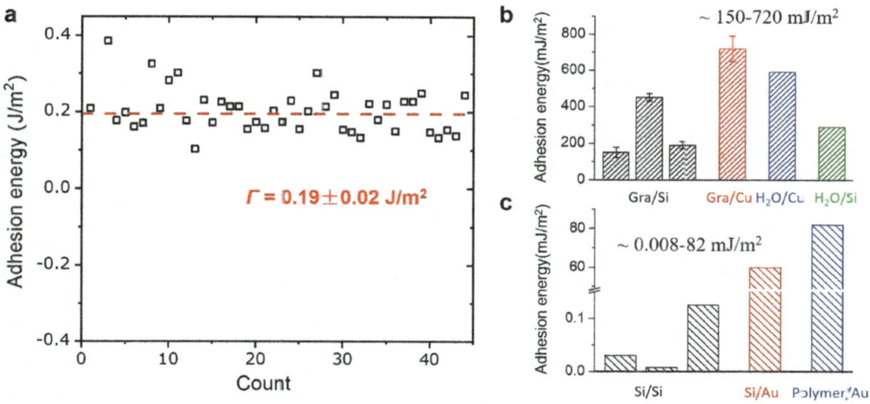

Fig. 4.12 **a** Experimentally measured interfacial adhesion energy between graphene and silicon substrate. **b** The interfacial adhesion energy between graphene and silicon substrate is comparable to the solid–liquid adhesion energies. **c** The solid–solid adhesion energies

regarding whether it will also modify the adhesion properties. Particularly, the coefficient C in Eq. (4.35) is related to b_0 that depends on the interfacial friction. We hence redefine the bubble volume (V_b)

$$
\begin{aligned}
V_b &= \int w(r)2\pi r dr = \int_0^a \left(\frac{\Delta p a^4}{Et}\right)^{\frac{1}{3}} \sum_{n=0}^{\infty} a_{2n}\left[1-\left(\frac{r}{a}\right)^{2n}\right]2\pi r dr \\
&= \left(\frac{\Delta p a^4}{Et}\right)^{\frac{1}{3}} \pi a^2 \left(a_0 + \sum_{n=0}^{\infty}\frac{a_{2n}}{n+1}\right) \\
&= C\pi a^2 \delta = C\pi a^2 \left(\frac{\Delta p r^4}{Et}\right)^{\frac{1}{3}} \sum_{n=0}^{\infty} a_{2n}
\end{aligned}
\tag{4.36}
$$

and obtain the expression $C = \frac{(a_0+\sum_{n=0}^{\infty}\frac{a_{2n}}{n+1})}{\sum_{n=0}^{\infty}a_{2n}}$ where $a_0 = \frac{1}{b_0}; a_2 = \frac{1}{2{b_0}^4}; a_4 = \frac{5}{9{b_0}^7}; a_6 = \frac{55}{72{b_0}^{10}} \ldots.$

We plot the $C - \bar{q}^{-\frac{2}{3}} / \bar{\tau}$ curve in Fig. 4.13, where the change of C is even below 5% with varying $\bar{q}^{-\frac{2}{3}} / \bar{\tau}$. Previous studies demonstrated that, even if C is increased by 10% from the value obtained from Hencky's solution, the resulting increase in the calculated averaged adhesion energy is only about 3.4% [18]. Hence, we speculate that the interfacial shear effect is negligible on the measurement of adhesion energy. However, wrinkling and radial buckle delamination have been widely observed during the bubble deformation, which is facilitated by sliding of the graphene membrane near

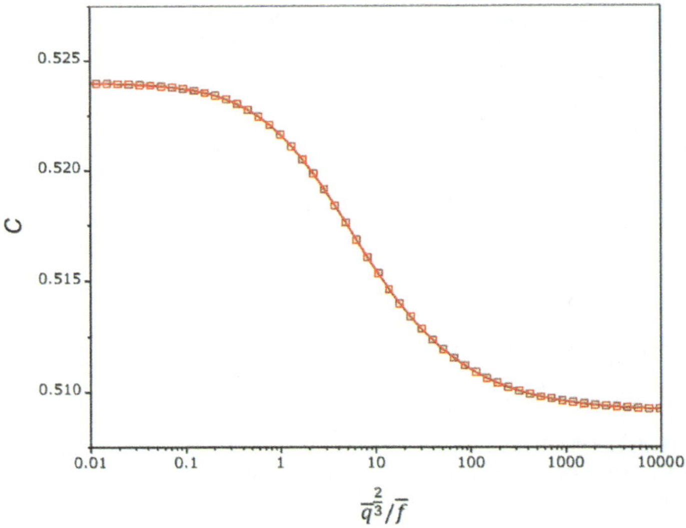

Fig. 4.13 The dependency of the parameter C on $\bar{q}^{-\frac{2}{3}} / \bar{\tau}$

the perimeter of the blister [18, 61, 62]. It is conceivable that such a surface instability would play a significant role in the delamination behavior of graphene in the substrate-supported region, especially under the cyclic loading.

Figure 4.14 illustrates the interfacial delamination of graphene bubbles at different locations within the same flake when the applied pressure (p_0) is as high as 1.5 MPa. Interestingly, distinct final morphologies are observed. In particular, the bubble located at the interior of the flake (#2) shows a uniform and symmetrical spherical cap shape as well as a smooth periphery, in consistency with the results reported

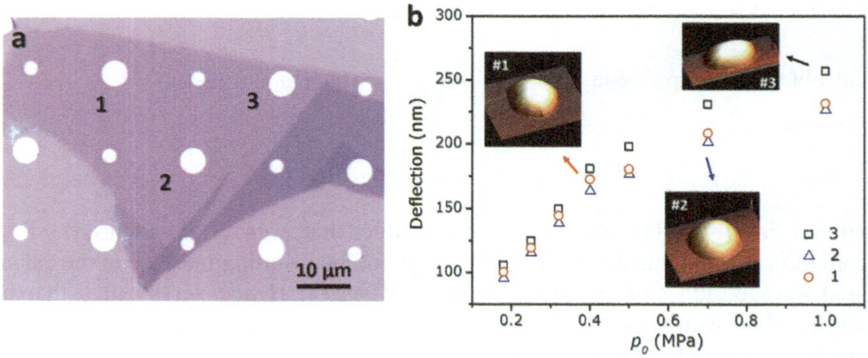

Fig. 4.14 a The optical image of the graphene flakes covered holes on silicon substrate. b Bubble deflection of graphene drums at three different positions as a function of applied pressure

in the previous literature [15, 18, 40]. By contrast, the bubble close to the graphene flake edge (#3) develops into a semi-ellipsoid shape, with its long axis parallel to the edge. Radial buckles can also be visible at two ends of long axis. As expected, the bubble (#1) sitting in between presents a polygonal shape with a few apparent buckles extending outward in varying degrees, corresponding to the transition from spherical cap to semi-ellipsoid. On the other hand, the evolution of the graphene bubble with increasing applied pressure was recorded by AFM in Fig. 4.14b. Whilst the deflections for three bubbles appear similar to each other for $p_0 < 0.4$ MPa, a deviation can be observed at higher applied pressure, yet still lower than critical pressure for delamination (~1.2 MPa). Specially, the deflection for bubble #3 reaches 257 nm at $p_0 = 1.0$ MPa, much larger than the theoretical prediction by Hencky's solution. According to the discussions in Sect. 4.4.2 (Fig. 4.9d), such a softening behavior is originated from the interfacial shear deformation in the annulus region outside the graphene bubbles as identified by the Raman mapping. Moreover, the inward sliding would compress the graphene circumferentially, leading to an elastic instability with respect to the formation of wrinkles, as evidenced in Fig. 4.15b. Further increasing pressures can induce radial buckles to release the hoop compression and gradually removes the round shape of graphene bubbles (Fig. 4.15c–f). Regarding the surface instabilities, Dai et al. indicated that the specific type depends on competition between friction and adhesion: When the interfacial interaction between the graphene membrane and the substrate is relatively weak tangentially but strong vertically, this contrast may result in the scenario where graphene wrinkles around the bubble border and remains flat in the supported region; however, when the substrate adhesion is weak enough to allow substantial interfacial sliding, radial buckles would

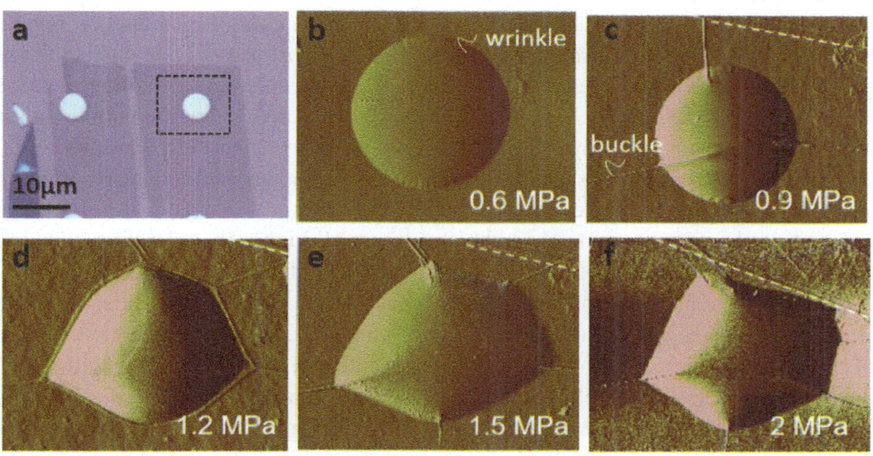

Fig. 4.15 **a** Optical image of a representative graphene sample and **b–e** evolution of bubble morphology with increasing applied pressures. The white dashed line is the reference to indicate the propagation of perimeter of bubble

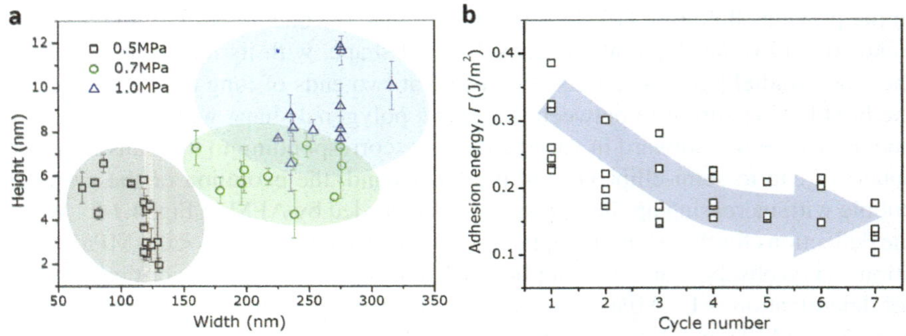

Fig. 4.16 **a** The evolution of buckles, including heights and widths, with increasing applied pressure. **b** The interfacial adhesion energy as a function of the loading cycle numbers

be activated [63]. Therefore, the emergence of radial buckles in Fig. 4.15c–f suggests a weakening of interface at higher pressures.

Note that, the evolution of surface morphology was recorded by the AFM measurement on the same graphene bubble, with each pressure corresponding to an inflation-deflation cycle. It hence recalls the interfacial degradation during cyclic loading as discussed in Sect. 2.4.1. As shown in Fig. 4.16a, the buckles were found to grow in size with increasing cycles, which would damage the full conformation of graphene to the substrate. The lowered contact quality is expected to reduce the adhesion energy, as evidenced in Fig. 4.16b. Obviously, the adhesion energy greatly decreases during the first three cycles, following with a plateau as the buckles stabilize in height and width in further cycles. However, our results seem to contradict with previous reports by Boddeti et al., where the apparent adhesion energy increases as bubble border propagates away from the cavity edge at high pressure. They attributed the lower apparent adhesion energy to the topographic and roughness variations near the perimeter of the cavity [18]. But one thing should be noted is that the bubble radius dependent adhesion was only observed for small bubbles with radius less than 2 μm, while absent for larger ones. In this case, the bubble deflection and corresponding strain level would be limited, which greatly suppresses the elastic instability. In contrast, in the present work, the bubble radii are all higher than 2.5 μm, allowing the formation of wrinkles/buckles that dominate the adhesion performance.

Having demonstrated the sliding boundary as the mechanism underlying the elastic instabilities, more attention was paid to the evolution of surface morphology of graphene bubbles. As shown in Fig. 4.15c–f, interestingly, the bubbles with radial buckles around tend to develop into irregular shapes at higher pressures under cyclic loading. The buckles act as the interfacial defects where the delamination preferably initiates and propagates [64]. To confirm the delamination mechanism led by buckles, we carried out the molecular dynamics (MD) simulations as illustrated in Fig. 4.17. The distance from the cavity to the flake edge are set different to reflect interfacial slip effect. Here, we define P_x as the distance from the cavity to the origin, so that a lower P_x is associated with a farther distance to flake edge. High

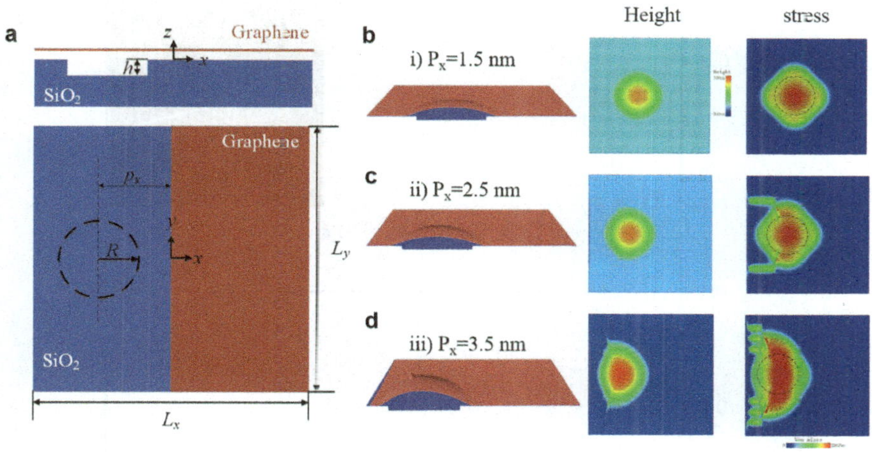

Fig. 4.17 **a** Modelled graphene/cavity/silicon substrate system in molecular dynamics simulations. **b** The distribution of height and stress in graphene bubble, which shows different distances from the hole center to the edge of (**a**) $P_x = 1.5$ nm, (**b**) $P_x = 2.5$ nm and (**c**) $P_x = 3.5$ nm

enough pressure load was applied to enable the delamination of the graphene from the substrate. For $P_x = 1.5$ nm, the delaminated graphene bubble presents a round shape without buckles and the corresponding stress distribution appears centrosymmetric (Fig. 4.17b). By comparison, as the cavity moves closer to the flake edge ($P_x = 2.5$ nm), buckles are generated upon the delamination and the bubble shape becomes non-round as shown in the height image. Accordingly, a nonuniform stress distribution is observed, featuring a stress concentration at the junction between the buckle and bubble (Fig. 4.17c). When the cavity is in close proximity to the flake edge ($P_x = 3.5$ nm), the bubble height shows a pronounced increase due to the significant interfacial sliding at the flake edge (Fig. 4.17d). In addition, the radial buckles grow in both the amount and size, resulting in a larger aspect ratio in geometry as well as a higher inhomogeneity of stress fields. More importantly, it is plausible that a closer distance from the cavity to the flake edge is likely to cause a broader angle between buckles, which determines the final morphology of the bubble. Following the evolutions in Fig. 4.17b–d, it can be envisaged that a semi-ellipsoid shape would be obtained with further increasing P_x, considering more buckles that are almost parallel to the flake edge.

As a matter of fact, such irregularly shaped bubbles have been observed between vdW materials trapping contamination (adsorbed water and hydrocarbons), which are spontaneously formed during the fabrication or mechanical transfer process [65, 66]. Though these bubbles are in nanoscale size, they share the mechanical mechanism behind the bubble geometry, namely, the aspect ratio (δ/R) is determined by the competition between vdW adhesion and elastic energies. Given that Eq. (4.35) fails to estimate the adhesion energy of irregular bubbles, Khestanova et al. proposed a universal scaling to describe all bubbles with different materials and shapes (e.g.

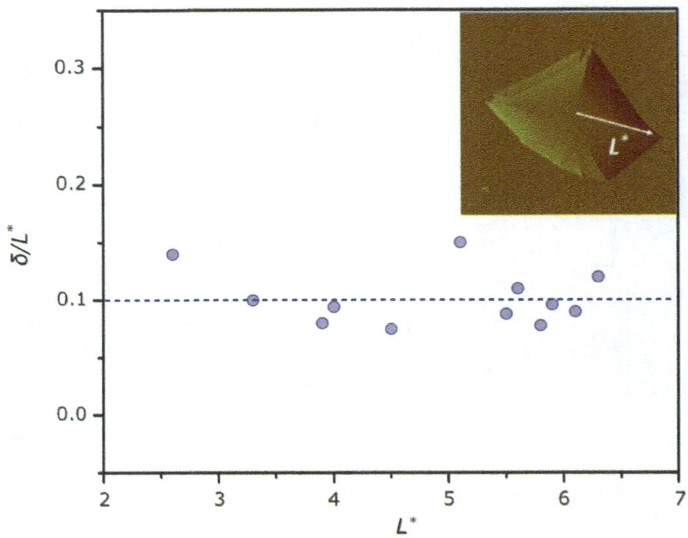

Fig. 4.18 The measured δ/L^* of polygonal bubbles, as a function of L^*

round, triangle). Specifically, δ/L^* was used for the polygonal bubble, where L^* is defined as the average distance from the center of flake to the polygon vertices to characterize the lateral dimension, as shown in the inset of Fig. 4.18. Then, the interfacial adhesion between different vdW materials can be compared in terms of the aspect ratio. Here we collect data of δ/L^* for 12 graphene samples, showing an average of 0.10. This value is within the range of δ/L^* (0.07–0.11) for hexagonal boron nitride (hBN) that has similar elastic properties, implying a comparable adhesion energy.

4.6 Chapter Summary

In combination of micro-Raman spectroscopy and AFM, we develop an experimental platform for characterizing vdW interfaces in graphene NEMS based on the pressurized bulging device. We uncover that the atomic smoothness and flexibility of the graphene render a robust vertical contact with silicon substrate but a considerably weak tangential resistance for the interface to shear deformations. As a consequence, the graphene membrane could be readily pulled to slide over its substrate by the vertical pressure load, even if the bubble deflection is low and the interfacial contact remains intact. In situ Raman imaging serves to monitor such interfacial shear deformations, which are activated in an annular zone growing controllably outwards. While the slip effect is found to cause the elastic softening of graphene bubble, the adhesion energy shows no clear dependence on it. However, the sliding boundary can lead to the formation of radial buckles, which act as interfacial defects to

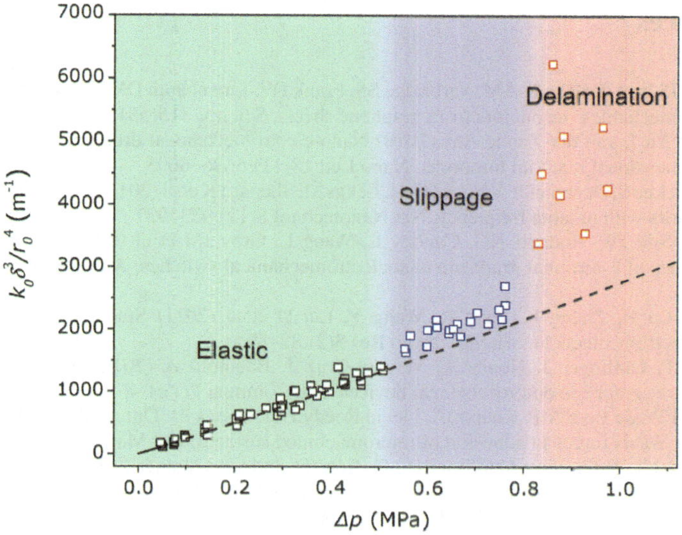

Fig. 4.19 Normalized center deflection $k_0\delta^3/r_0{}^4$ versus the applied pressure difference shows a three-stage feature, including elastic stage, sliding stage and delamination stage

accelerate the delamination and define the final shape of bubble. Though key parameters remain to be determined to quantify such connections between sliding-buckle and delamination-bubble, the current understanding can be exploited to design and control the bubble/buckle network geometry through interface engineering.

Furthermore, the combined Raman-AFM-bulging technique can be extended to other vdW materials, such as hBN, transition metal dichalcogenide (TMD), black phosphorus and Mxene, offering a comprehensive view in both intrinsic and interfacial mechanical properties. As presented in Fig. 4.19, the bubble deformation is typically divided into three different stages: (1) Initially, the bubble elastically deforms with a fully clamped boundary. Based on the AFM measurement of the deflections as a function of pressures, elastic properties can be estimated, including Young's modulus, bending rigidity and Poisson's ratio; (2) As the pressure load increases, the radial stress generated in bubbles is prone to pull the supported graphene membrane inwards and hence induce the interfacial slippage. By capturing the controllable growth of interfacial shear zone using Raman mapping technique, the interfacial shear stress can be obtained based on a membrane analysis; (3) When the critical pressure for delamination is reached, the graphene membrane stably detaches from the substrate, allowing the evaluation of adhesion energy by measuring the bubble height and radius with AFM. Even though the bubble develops into a non-round shape due to the presence of radial buckles, the universal scaling behavior still enables a comparison of adhesion property between different materials.

References

1. Bunch JS, van der Zande AM, Verbridge SS, Frank IW, Tanenbaum DM, Parpia JM et al (2007) Electromechanical resonators from graphene sheets. Science 315(5811):490–493
2. Kim S, Yu J, van der Zande AM (2018) Nano-electromechanical drumhead resonators from two-dimensional material bimorphs. Nano Lett 18(11):6686–6695
3. Chen C, Lee S, Deshpande VV, Lee G-H, Lekas M, Shepard K et al (2013) Graphene mechanical oscillators with tunable frequency. Nat Nanotechnol 8(12):923–927
4. Liu X, Suk JW, Boddeti NG, Cantley L, Wang L, Gray JM et al (2014) Large arrays and properties of 3-terminal graphene nanoelectromechanical switches. Adv Mater 26(10):1571–1576
5. Shi Z, Lu H, Zhang L, Yang R, Wang Y, Liu D et al (2011) Studies of graphene-based nanoelectromechanical switches. Nano Res 5(2):82–87
6. Weber P, Güttinger J, Noury A, Vergara-Cruz J, Bachtold A (2016) Force sensitivity of multilayer graphene optomechanical devices. Nat Commun 7(1):1–8
7. Delrio FW, de Boer MP, Knapp JA, David Reedy E Jr, Clews PJ, Dunn ML (2005) The role of van der Waals forces in adhesion of micromachined surfaces. Nat Mater 4(8):629–634
8. Schneeweiss P, Gierling M, Visanescu G, Kern D, Judd T, Günther A et al (2012) Dispersion forces between ultracold atoms and a carbon nanotube. Nat Nanotechnol 7(8):515–519
9. Israelachvili J, Tabor D (1972) The measurement of van der Waals dispersion forces in the range 1.5 to 130 nm. Proc R Soc Lond A Math Phys Eng Sci R Soc, 19–38.
10. Sarabadani J, Naji A, Asgari R, Podgornik R (2011) Many-body effects in the van der Waals-Casimir interaction between graphene layers. Phys Rev B 84(15):155407
11. Dai Z, Liu L, Zhang Z (2019) Strain engineering of 2D materials: issues and opportunities at the interface. Adv Mater 31(45):1805417
12. Megra YT, Suk JW (2019) Adhesion properties of 2D materials. J Phys D Appl Phys 52(36):364002
13. Bunch JS, Dunn ML (2012) Adhesion mechanics of graphene membranes. Solid State Commun 152(15):1359–1364
14. Zong Z, Chen C-L, Dokmeci MR, Wan K-T (2010) Direct measurement of graphene adhesion on silicon surface by intercalation of nanoparticles. J Appl Phys 107(2):026104
15. Koenig SP, Boddeti NG, Dunn ML, Bunch JS (2011) Ultrastrong adhesion of graphene membranes. Nat Nanotechnol 6(9):543–546
16. Cao Z, Wang P, Gao W, Tao L, Suk JW, Ruoff RS et al (2014) A blister test for interfacial adhesion of large-scale transferred graphene. Carbon 69:390–400
17. Xin H, Borduin R, Jiang W, Liechti KM, Li W (2017) Adhesion energy of as-grown graphene on copper foil with a blister test. Carbon 123:243–249
18. Boddeti NG, Liu X, Long R, Xiao J, Bunch JS, Dunn ML (2013) Graphene blisters with switchable shapes controlled by pressure and adhesion. Nano Lett 13(12):6216–6221
19. Zhang K, Arroyo M (2014) Understanding and strain-engineering wrinkle networks in supported graphene through simulations. J Mech Phys Solids 72:61–74
20. Zhang K, Arroyo M (2013) Adhesion and friction control localized folding in supported graphene. J Appl Phys 113(19):193501
21. Kitt AL, Qi Z, Remi S, Park HS, Swan AK, Goldberg BB (2013) How graphene slides: measurement and theory of strain-dependent frictional forces between graphene and SiO_2. Nano Lett 13(6):2605–2610
22. Wang G, Dai Z, Wang Y, Tan P, Liu L, Xu Z et al (2017) Measuring interlayer shear stress in bilayer graphene. Phys Rev Lett 119(3):036101
23. Novoselov KS, Jiang D, Schedin F, Booth TJ, Khotkevich VV, Morozov SV, Geim AK (2005) Two-dimensional atomic crystals. Proc Natl Acad Sci 102(30):10451–10453
24. Fichter WB (1997) Some solutions for the large deflections of uniformly loaded circular membranes. NASA technical paper 3658, pp 1–24
25. Peng W, Gao W, Cao Z, Liechti KM, Huang R (2013) Numerical analysis of circular graphene bubbles. J Appl Mech 80:040905

26. Timoshenko SP, Goodier J (2011) Theory of elasticity. Int J Bulk Solids Storage in Silos 1(4):567–567
27. Hencky H (1915) On the stress state in circular plates with vanishing bending stiffness. Z Math Phys 63:311–317
28. Radha B, Esfandiar A, Wang F, Rooney A, Gopinadhan K, Keerthi A et al (2016) Molecular transport through capillaries made with atomic-scale precision. Nature 538(7624):222–225
29. Yue K, Gao W, Huang R, Liechti KM (2012) Analytical methods for the mechanics of graphene bubbles. J Appl Phys 112(8):083512
30. Boddeti NG, Koenig SP, Long R, Xiao J, Bunch JS, Dunn ML (2013) Mechanics of adhered, pressurized graphene blisters. J Appl Mech 80(4):040909
31. Bunch JS, Verbridge SS, Alden JS, van der Zande AM, Parpia JM, Craighead HG et al (2008) Impermeable atomic membranes from graphene sheets. Nano Lett 8(8):2458–2462
32. Sun PZ, Yang Q, Kuang WJ, Stebunov YV, Xiong WQ, Yu J et al (2020) Limits on gas impermeability of graphene. Nature 579(7798):229–232
33. Neumann C, Reichardt S, Venezuela P, Drogeler M, Banszerus L, Schmitz M et al (2015) Raman spectroscopy as probe of nanometre-scale strain variations in graphene. Nat Commun 6:8429
34. Metzger C, Rémi S, Liu M, Kusminskiy SV, Castro Neto AH, Swan AK et al (2009) Biaxial strain in graphene adhered to shallow depressions. Nano Lett 10(1):6–10
35. Ferrari AC, Basko DM (2013) Raman spectroscopy as a versatile tool for studying the properties of graphene. Nat Nanotechnol 8(4):235–246
36. Frank O, Tsoukleri G, Riaz I, Papagelis K, Parthenios J, Ferrari AC et al (2011) Development of a universal stress sensor for graphene and carbon fibres. Nat Commun 2:255
37. Dai Z, Hou Y, Sanchez DA, Wang G, Brennan CJ, Zhang Z et al (2018) Interface-governed deformation of nanobubbles and nanotents formed by two-dimensional materials. Phys Rev Lett 121(26):266101
38. Lloyd D, Liu X, Boddeti N, Cantley L, Long R, Dunn ML et al (2017) Adhesion, stiffness, and instability in atomically thin MoS_2 bubbles. Nano Lett 17(9):5329–5334
39. Dai Z, Wang G, Liu L, Hou Y, Wei Y, Zhang Z (2016) Mechanical behavior and properties of hydrogen bonded graphene/polymer nano-interfaces. Compos Sci Technol 136:1–9
40. Liu X, Boddeti NG, Szpunar MR, Wang L, Rodriguez MA, Long R et al (2013) Observation of pull-in instability in graphene membranes under interfacial forces. Nano Lett 13(5):2309–2313
41. Metten D, Federspiel F, Romeo M, Berciaud S (2014) All-optical blister test of suspended graphene using micro-raman spectroscopy. Phys Rev Appl 2(5):054008
42. Zhang T, Li X, Gao H (2015) Fracture of graphene: a review. Int J Fract 196(1–2):1–31
43. Kosevich A, Lifshitz E, Landau L, Pitaevskii L (1986) Theory of elasticity. Theoret Phys 7.
44. Wang X, Tantiwanichapan K, Christopher JW, Paiella R, Swan AK (2015) Uniaxial strain redistribution in corrugated graphene: clamping, sliding, friction, and 2D band splitting. Nano Lett 15(9):5969–5975
45. Young RJ, Kinloch IA, Gong L, Novoselov KS (2012) The mechanics of graphene nanocomposites: a review. Compos Sci Technol 72(12):1459–1476
46. Gong L, Kinloch IA, Young RJ, Riaz I, Jalil R, Novoselov KS (2010) Interfacial stress transfer in a graphene monolayer nanocomposite. Adv Mater 22(24):2694–2697
47. Papageorgiou DG, Kinloch IA, Young RJ (2017) Mechanical properties of graphene and graphene-based nanocomposites. Prog Mater Sci 90:75–127
48. Sanchez DA, Dai Z, Wang P, Cantu-Chavez A, Brennan CJ, Huang R et al (2018) Mechanics of spontaneously formed nanoblisters trapped by transferred 2D crystals. Proc Natl Acad Sci 115(31):7884–7889
49. Jensen HM (1991) The blister test for interface toughness measurement. Eng Fract Mech 40(3):475–486
50. Chen J, Bull SJ (2011) Approaches to investigate delamination and interfacial toughness in coated systems: an overview. J Phys D Appl Phys 44(3):034001
51. Williams JG (1997) Energy release rates for the peeling of flexible membranes and the analysis of blister tests. Int J Fract 87:265–288

52. Kim J, Kim K, Kim Y (1989) Mechanical effects in peel adhesion test. J Adhes Sci Technol 3(1):175–187
53. Carpick RW, Salmeron M (1997) Scratching the surface: fundamental investigations of tribology with atomic force microscopy. Chem Rev 97(4):1163–1194
54. Zhang S, Sun D, Fu Y, Du H (2005) Toughness measurement of thin films: a critical review. Surf Coat Technol 198(1–3):74–84
55. Gent A, Lewandowski L (1987) Blow-off pressures for adhering layers. J Appl Polym Sci 33(5):1567–1577
56. Israelachvili JN (2011) Intermolecular and surface forces. Academic Press
57. Maboudian R (1997) Critical review: adhesion in surface micromechanical structures. J Vac Sci Technol 15(1):1–20
58. Buks E, Roukes M (2001) Stiction, adhesion energy, and the Casimir effect in micromechanical systems. Phys Rev B 63(3):033402
59. DelRio FW, Dunn ML, Boyce BL, Corwin AD, De Boer MP (2006) The effect of nanoparticles on rough surface adhesion. J Appl Phys 99(10):104304
60. DelRio FW, Jaye C, Fischer DA, Cook RF (2009) Elastic and adhesive properties of alkanethiol self-assembled monolayers on gold. Appl Phys Lett 94(13):131909
61. Dai Z, Sanchez DA, Brennan CJ, Lu N (2020) Radial buckle delamination around 2D material tents. J Mech Phys Solids 137:103843
62. Ma Y, Wang G, Chen Y, Long D, Guan Y, Liu L et al (2018) Extended Hencky solution for the blister test of nanomembrane. Extreme Mech Lett 22:69–78
63. Dai Z (2020) Mechanics of bubbles and tents formed by 2D materials. The University of Texas at Austin, Austin
64. Ma Y, Wang G, Chen Y, Liu L, Zhang Z (2019) A mock gas molecules model for accurately simulating pressure load at micro- and nanoscales. J Appl Mech 86(9):091006
65. Khestanova E, Guinea F, Fumagalli L, Geim AK, Grigorieva IV (2016) Universal shape and pressure inside bubbles appearing in van der Waals heterostructures. Nat Commun 7:12587–12596
66. Ghorbanfekr-Kalashami H, Vasu K, Nair R, Peeters FM, Neek-Amal M (2017) Dependence of the shape of graphene nanobubbles on trapped substance. Nat Commun 8:15844–15854

Chapter 5
Interfacial Mechanics Between van der Waals Materials

5.1 Introduction

Two-dimensional (2D) materials, monolayer atoms with crystalline structures and unusual physical characteristics, have been fueling extensive research on their use in semiconductors, electronics, battery energy and composites [1–3]. For most exciting applications of 2D crystals, the emergence of multilayer structures is inevitable either because of the production process limitations [4–6], or more importantly, because of the functional and operating requirements [7–10]. Indeed, ordered structures comprised of two or more layers represent a wide class of materials and their extraordinary properties can be modulated by taking advantage of interlayer interactions. For instance, layered stacking of different 2D materials, generally called van der Waals (vdW) heterostructures, reveals exceptionally electronic and photonic properties [11–14]. The presence of interlayer gallery is also essential for tuning the frequencies a of mechanical resonators [7], improving thermal conductivity of nanocomposites [8], offering nanochannels for selective molecular permeation [9] and electrochemically charging for actuation, sensor and energy storage [15–17]. From classic view of multilayer system, one of the most critical issues that determine its overall performance, durability and even fabrication process, lies in the interfacial mechanics [18]. Moreover, the interface effects could be much more critical for the atomically thin layers that fundamentally thermal, electronic, optical and tribological properties could be affected by the interlayer deformation (especially under shear mode) of multilayer structures [12, 19–21], whereas corresponding mechanical parameters are far from well-characterized.

Significant developments have been achieved in recent years in characterizing the mechanics of graphene and other 2D crystals beyond graphene. The multilayer system was typically treated as a single sheet by which the interlayer deformation is often overlooked [22–24]. Such treatment also assumed strong interaction between layers, making the exploration on interlayer shear parameters elusive. In fact, the interlayer vdW interactions for most 2D materials are relatively weak in the mechanical sense so that individual layers with atomically smooth surfaces are highly

G. Wang, *Characterization and Modification of Graphene-Based Interfacial Mechanical Behavior*, Springer Theses, https://doi.org/10.1007/978-981-15-8029-1_5

lubricated [25–27]. Recently, the interlayer shear behavior has also been explored through recording the interaction between multilayer nanosheets and scanning tip via frictional force microscopy [28–30] or indentation tip via atomic force microscope (AFM) [31, 32]. Although these tip-based measurements with stress concentrations provided interesting dynamic phenomenon, the understanding of deformation mechanisms in multilayer structures remains experimentally challenging and the quantitative characterization of the interlayer interaction has yet to emerge.

The possibility of tailoring the chemistry of graphene emerged as a powerful tool to endow this material with intriguing properties or to modify on demand its intrinsic capabilities [33–35]. In addition to the in-plane mechanical properties, it is fundamentally important to understand how the interlayer interaction of neighboring graphene layers is influenced by chemical impurities. Monitoring chemically doped interfacial coupling in multilayer graphene is therefore essential both from a fundamental interest in understanding its atom-level interlayer interactions and from a technological perspective for its reliable applications.

To resolve these issues, we employ the Raman-AFM-bulging technique proposed in the last chapter to characterize the bubble deformation of bilayer graphene and monitor the interlayer shear deformation activated in the shear zone. Following the membrane analysis, the interlayer shear stress between graphene layers is measured for the first time. The effect of the interlayer shear on the bending performance of 2D materials is further elucidated, addressing the puzzle as to how the bending rigidity of a multilayer 2D material relates to its layer number. Finally, the boron doping is implemented to tailor the interlayer coupling of multilayer graphene, which opens an avenue for other heteroatoms such as nitrogen, sulfur, silicon and phosphorous [36–38].

5.2 Experimental Methods

5.2.1 Preparation and Characterization of van der Waals Materials

Two-dimensional (2D) materials including graphene, hexagonal boron nitride (hBN) and molybdenum disulfide (MoS_2), were prepared on pre-patterned silicon substrates covered with a 300 nm thick SiO_2 layer by mechanical exfoliation. Atomic force microscopy (AFM) offers a direct technique to identify the thickness of 2D materials. Take graphene as an example. Figure 5.1a shows the optical image of a graphene flake containing multiple layers on the SiO_2/Si substrate. Figure 5.1b presents the AFM image of the rectangular area highlighted in Fig. 5.1a, where the tapping mode was employed to minimize possible mechanical damages. The thickness measurements were carried out by two line-scans, and the corresponding values are indicated in the inset of Fig. 5.1b. Although the thickness of the monolayer flake may be different possibly due to the instrumental offset [39, 40], the thickness difference

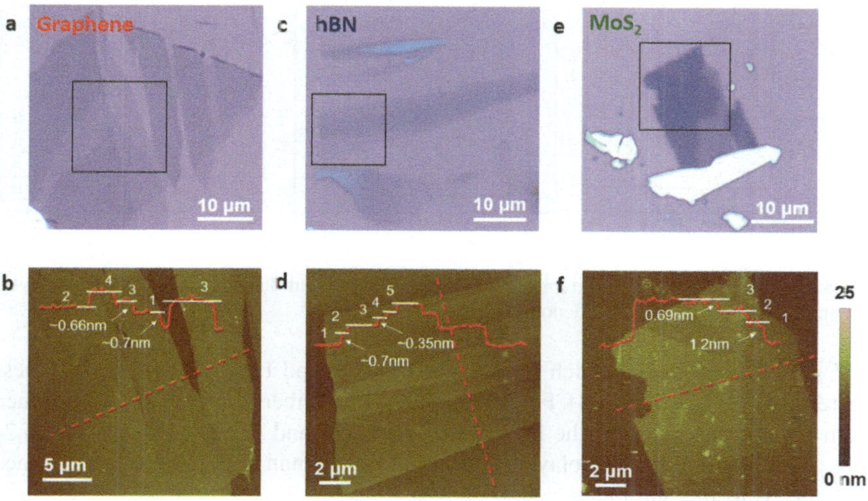

Fig. 5.1 Optical images of **a** graphene, **c** hBN and **e** MoS_2 sheets with various layer numbers. The magnified views of the rectangular regions are shown in AFM images for **b** graphene, **d** hBN and **f** MoS_2, respectively. The insets give the height profiles along the red dashed lines, where the interlayer distance was found to be around 0.34 nm, 0.35 nm and 0.69 nm for graphene, hBN, and MoS_2, respectively. The relatively high thickness of the bottom layer (0.7 nm for graphene and hBN, and 1.2 nm for MoS_2) was typically attributed to the instrumental scanning offset from the substrate to the 2D materials

due to different numbers of layers stacked together gave a quite consistent interlayer distance ($t_1 = 0.34$ nm). Similar phenomena were also observed for hBN and MoS_2 as shown in Fig. 5.1d, f, whose interlayer distances were measured to be ~0.35 nm and ~0.69 nm, respectively. Therefore, a simple linear relationship between the number of layers (N) and total measured thickness (t_{measured}) can be obtained as

$$t_{\text{measured}} = s + Nt_1 \tag{5.1}$$

where s is the substrate-caused offset for the bottom layer (0.36 nm for graphene, 0.35 nm for hBN, and 0.51 nm for MoS_2). The layer numbers of various van der Waals materials are hence determined by the AFM measurements, with an estimated error less than 5%.

To further calibrate the AFM measurements, Raman spectroscopy was also utilized to determine the thickness of few-layer 2D materials ($N \leq 4$) [41]. As shown in Fig. 5.2a, the 2D profiles of graphene exhibit distinct features with the layer number increasing from 1 to 4, while higher thickness can be discriminated by the enhanced G peak intensity (E_{2g} mode). In detail, it has been demonstrated that the intensity ratio between the G-peak and Si-peak from bare SiO_2/Si substrates can be used for determining the layer number of graphene flakes up to 100 layers [42]. In contrast, Raman spectra of hBN sheets show only a Raman G-band corresponding

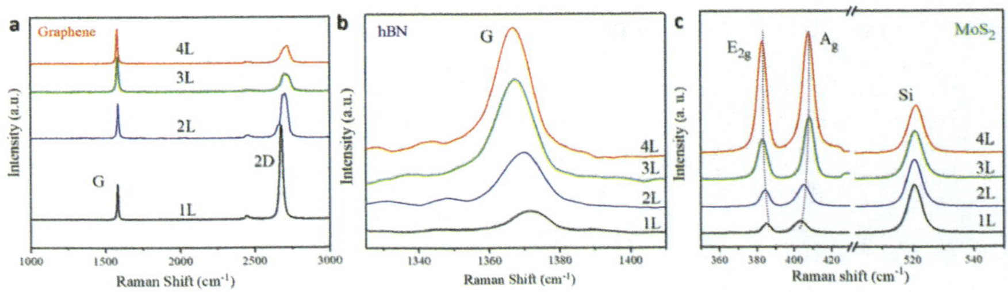

Fig. 5.2 Raman spectrum for **a** graphene, **b** hBN and **c** MoS_2 multilayers with the number of layers ranging from 1 to 4. Here 'NL' denotes a N-layer sheet

to E_{2g} vibration mode, which commonly upshifts and broadens as the thickness decreases (Fig. 5.2b) [43, 44]. For MoS_2, the layer numbers are typically determined from the distance ($\Delta\omega$) of the Raman modes of E_{2g} and A_{1g} as shown in Fig. 5.2c (e.g. $\Delta\omega$=18 cm^{-1} for monolayers) [45, 46]. The Raman results were well aligned with the AFM measurements.

5.2.2 *Boron Doping*

The boron doping process was achieved by mixing boron oxide with natural graphite and thermally treating the mixture at 2200 °C for about 2 h using a graphite furnace in an argon atmosphere [47]. The weight ratio of boron oxide to graphite ranged from 0.5 to 50. The doping type and the boron content in the graphite were evaluated using X-ray photoelectron spectroscopy (XPS, ESCALAB 250Xi). The powder diffraction patterns of graphite before and after boron doping were obtained using synchrotron X-ray diffraction (XRD, Rigaku D/max-2500). The high-frequency Raman spectra of graphite and exfoliated graphene on SiO_2/Si substrate were measured with a Jobin–Yvon HR800 Raman system, equipped with a liquid nitrogen cooled charge-coupled detector. The laser excitation wavelengths are selected as 532 nm from a diode-pumped solid-state laser. High-resolution transmission electron microscopy (HRTEM) instruments (FEI Tecnai G2 F20 U-TWIN) were used for high-resolution imaging. Scanning tunneling microscopy (STM) experiments were carried out with a Nanoscope IIIa scanning probe microscope system (Veeco Metrology, USA) in ambient conditions.

5.2.3 *Low-Frequency Raman Shear Mode*

Low-frequency Raman measurements were performed in a backscattering geometry at room temperature using a Jobin–Yvon HR800 Raman system. The excitation wavelength is 532 nm of a diode pumped solid-state laser. Plasma lines are removed from the laser signals using BragGrate Bandpass filters (OptiGrate Corp). The Rayleigh line is suppressed by three BragGrate notch filters at 532 nm with optical density 3 and with the spectral bandwidth ~5–10 cm^{-1}, enabling the measurements of Raman modes down to 5 cm^{-1} [20]. All the Raman measurements were performed using a 100× objective lens (NA = 0.90) and a 1800 lines/mm grating to get a spectral resolution 0.35 cm^{-1} per CCD pixel. The typical laser power is ~0.5 mW to avoid sample heating. The accumulation time for each spectrum is ~600 s.

5.2.4 AFM Nanoindentation Test

Suspended graphene flakes were prepared on pre-patterned silicon substrates covered with a 90 nm thick SiO_2 layer by micromechanical cleavage (Fig. 5.3a). An array of circular holes was fabricated by photolithography and reactive ion etching. The depth is ~1 μm and the diameter ranges from 0.5 to 1.5 μm. The nanoindentation test on the suspended graphene was performed using a commercial AFM system (Asylum Research, MFP-3D Infinity). Before each indentation, the samples were scanned in tapping mode to find graphene sheets fully covering a hole. After scanning, the AFM tip (NanoScience Instruments) was centered in the middle of the circular hole. Mechanical testing was performed using force–displacement mode. The force versus displacement curves obtained from the AFM nanoindentation tests were used to quantitatively determine the elastic stiffness of boron-doped graphene and the observed hysteresis loops enabled the estimation of the energy dissipation due to the possible slippage.

Note that, previous attempts to obtain reproducible results with standard silicon or slender diamond tips failed because the high stiffness of graphene membranes produced tip breaking [32, 48]. To solve this issue, as well as to have a constant and well-defined contact geometry, we use commercial tips from NanoScience Instruments with hemispherical geometry and low wear coating of Tungsten carbide with

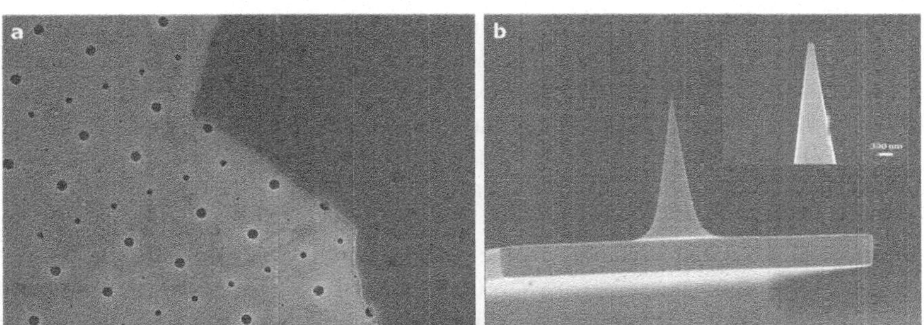

Fig. 5.3 **a** SEM images of exfoliated multilayer graphene sheets suspended over patterned holes. **b** SEM image of the AFM tip used for the measurements

nominal final tip radius of 60 nm (Fig. 5.3b) [49]. The cantilever spring constant was calibrated by Sader method before every series of testing, ranging from 38 to 40 N/m. All the measurements presented here were obtained at the same loading/unloading rate (300 nm/s). We follow the well-developed strategy of Lee's work to identify the zero displacement point or zero force level [48].

5.3 Interfacial Shear Between Bilayer Graphene

5.3.1 Pressurized Bulging Test

As Raman-AFM-bulging technique has been demonstrated to be versatile to access the interfacial properties of graphene, here we utilize it to explore the interlayer mechanics between bilayer graphene. Following the procedures in Sect. 4.2.1, suspended graphene sheets were prepared on pre-patterned silicon substrates covered with a 300 nm thick SiO_2 layer by micromechanical cleavage. An array of round holes was fabricated by photolithography and reactive ion etching. The depth was 300 nm and the diameters were 5 and 3 μm. The lateral dimension of an individual graphene sheet was identified by the optical microscopy. The vertical displacement and local strain in bubbles were measured by AFM and Raman spectroscopy, respectively. More details can be found in Sect. 4.2.2.

5.3.2 Propagation of Shear Zone Surrounding the Bubble

Figure 5.4a shows the Raman spectra recorded on bilayer graphene sheets for both the suspended and supported regions at different applied pressures. At $\Delta p = 0$ MPa, less pronounced differences can be observed in Raman G-band between suspended region and supported region, compared with that for monolayers. This is because the bilayer graphene has higher bending rigidity to resist the out-of-plane deformation, so that the initial strain is lower. As Δp increases to 0.28 MPa, commensurable downshifts can be observed in Raman spectrum of bilayer graphene bulging devices. In addition, the shape of 2D-mode of bilayer graphene tends to be almost symmetric, possibly implying that the Bernal (AB) stacking becomes lost possibly caused by the interlayer sliding [50]. When Δp reaches 0.69 MPa, the Raman shift is as high as 53 cm^{-1} (Fig. 5.4c). In addition, as the supported bilayer graphene membrane suffers from the radial stretching by the bubble, significant downshifts can be seen in supported region at Δp increases to 0.28 MPa, which is even larger than that of monolayer graphene. To gain more insights into the Raman characteristics of the supported graphene, we performed line-scanning along the diameter of the bubble, and a series of Raman spectra of G-band are stacked vertically as shown in Fig. 5.4b. It is evident that substantial variations in both the peak position and the full width at

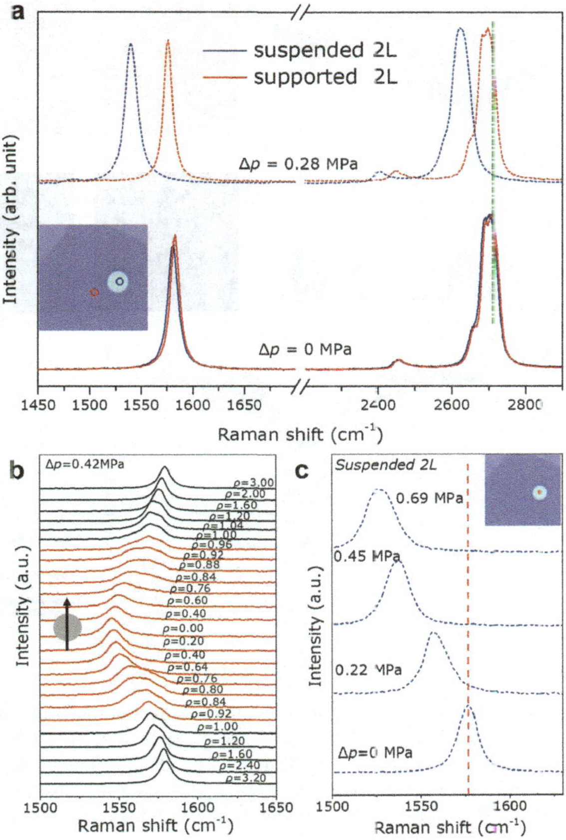

Fig. 5.4 **a** Raman spectra recorded on the bilayer (2L) bulging devices both for the suspended (blue line) and supported regions (red line) at different pressures. The optical image in the inset shows the locations at which Raman measurements are taken. **b** For a bilayer graphene bubble at Δp=0.42 MPa a series of Raman spectra (G-band) are stacked vertically in the direction of the line scan. **c** Raman spectra recorded in the center of suspended region on the monolayer graphene at different pressures

half-maximum (FWHM) can be detected. As the recorded position approaches to the bubble border ($\rho = 1.00$), an apparent peak broadening and splitting are highlighted due to the presence of high strain gradient. Even extended outside the cavity, the Raman G-band still exhibits a blueshift with increasing distance from the bubble, up to three times the radius of cavity. This actually suggests a faster growth of interfacial shear zone for bilayer graphene compared with monolayers, as illustrated in Fig. 5.5a.

Here, similar growth behavior of a shear zone outside the cavity was captured by the Raman mapping of G-band frequency (Fig. 5.5c). Compared to monolayer graphene, bilayer graphene requires a larger pressure difference to bulge up to the same maximum deflection, yielding a similar strain distribution inside the hole but a much larger shear zone outside the hole. These results suggest that for bilayer graphene bubbles, significant interlayer shear deformation occurred outside the hole. As illustrated in Fig. 5.5b, the interlayer shear zone would be much larger than the interfacial shear zone for the monolayer graphene if the interlayer shear stress, τ_2, between the two graphene layers is much smaller than τ_1 [51]. In other words,

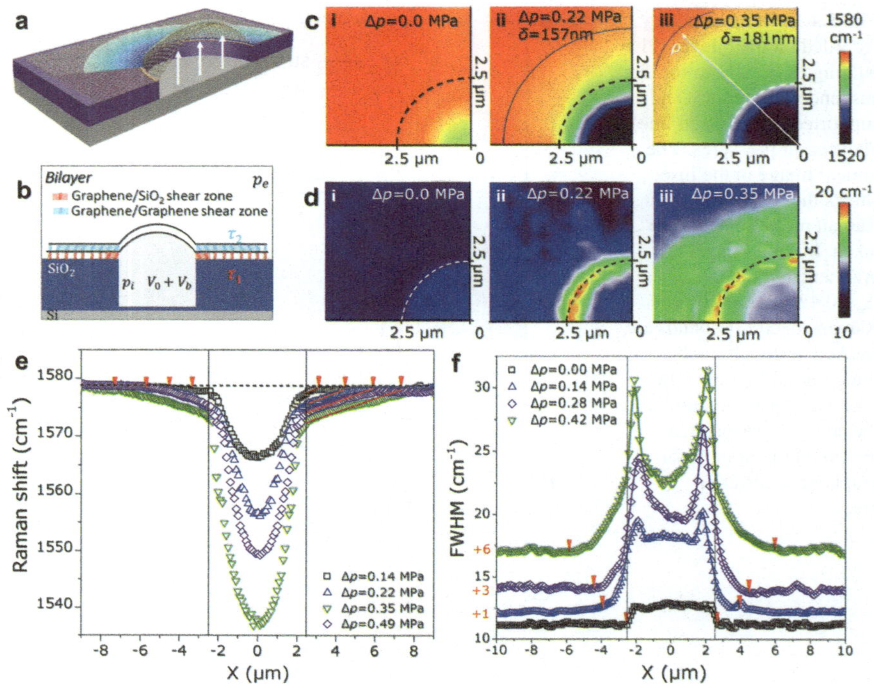

Fig. 5.5 **a** Schematic diagrams of bulged bilayer graphene samples. **b** The graphene-SiO_2 interface is dominated by τ_1 while the graphene-graphene interface is dominated by τ_2. Spatial maps of the **c** frequency and **d** width of the Raman G-band in the first quadrant for a bilayer graphene at different pressures, including suspended and supported regions discriminated by the dashed line. The cross sections of the contour mapping of Raman G-band **e** frequency and **f** width for the bilayer graphene membrane at different pressures

the much more discernable growth behavior or the larger shear zone observed in bilayer samples outside the hole is dominated by the shear deformation at the weak graphene/graphene interface rather than the graphene/substrate interface, which also strongly agrees with our Raman mapping results of the G-peak FWHM, as displayed in Fig. 5.5d. Furthermore, the cross sections of the contour mapping of Raman G-band frequency was also extracted as shown in Fig. 5.5e, showing a clear linearity near the edge of the shear zone and allowing the determination of ρ. In principle, the biaxial strain and stress in graphene may also be determined from the measured Raman shifts for bilayer graphene samples, which could then be used to determine the shear stresses for graphene/graphene interface. However, a precise relationship between the Raman shift and the biaxial strain of graphene (e.g. the Grüneisen parameter and shear deformation potential) is still challenging to be established when the interlayer coupling exists for bilayer graphene [52]. This issue might be resolved if independent Raman signals could be obtained for both bottom and top layers [53].

It has been reported that Raman peak width depends on the lattice deformation and is regarded as a measure of nanometer-scale strain variations in graphene [54]. In this regard, the FWHM gives us the invaluable capability of comprehensively probing the strain distribution in the case of graphene bubbles. Herein, line scans across the graphene bubbles, which encode the Raman G-band FWHM extracted from a single Lorentzian fit, are plotted in Fig. 5.5f with increasing applied pressures. Typically, the FWHM of bilayer graphene is around 11 cm^{-1} and it is broadened slightly over the hole at $\Delta p' = 0$ due to the absence of the charge transfer doping as well as the strain effect as discussed above. At $\Delta p = 0.14$ MPa, the formation of the graphene bubble clearly enhances the FWHM of G-band, especially at the bubble border. The overshoot of the G-band FWHM is likely a result of convoluting shifted peaks from the almost unstrained and highly strained regions as the diffraction limited focal spot is scanned across the bubble edge. These two signal components add up, yielding a non-Lorentzian shape with greater width [55]. Another factor worthy of close scrutiny is the G-band splitting induced by larger discrepancy between radial and tangential strains at the edge, which might also cause the peak broadening. With the pressure further rising, higher graphene bubbles create more complicated strain distributions around the edge, with visible fluctuations in Fig. 5.4b. We note that, as expected, the same features in the growth of the shear zone can be identified by both the G-band FWHM and frequency maps. The substantial enhancement of FWHM in the supported region of graphene is a clear indication of the biaxial strain as a result of the interlayer shear deformation, further corroborating our discussion of the growth of shear zones for bilayer graphene bubbles. Note that, in previous works [15, 33], pressurized blister tests were commonly performed to measure the adhesion energy between thin films and substrates as interfaces would delaminate at a critical pressure. Here, however, the interlayer shear deformation could be actuated under relatively low pressures (far from that required to cause delamination), due to the weak interlayer shear resistances.

As Raman shifts cannot provide direct evidence of the interlayer deformations at atomic-scale resolution, where the mechanism of interlayer load transfer through shear could be revealed, we carried out molecular dynamics (MD) simulations for the bulging tests. The results depicted in Fig. 5.6a, b clearly demonstrate the binding energy and radial displacement of a bilayer graphene sheet, indicating that there is no

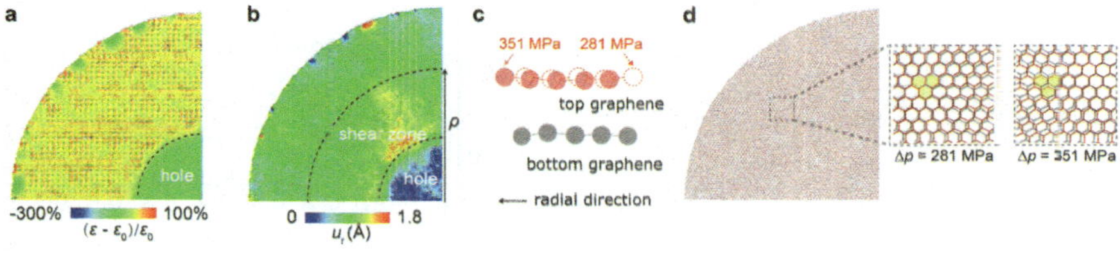

Fig. 5.6 **a** Spatial distribution of the interfacial binding energy between bottom-layer graphene and the silica substrate. **b** Decomposed in-plane displacement u_r in the top-layer graphene calculated from the MD simulation results. **c** The interlayer shear behaviors of bottom-layer and top-layer graphene, resolved by **d** the atomic displacement under specific pressure difference

debonding in our MD simulations in consistence with the experiments. The interlayer shear deformation can be visualized from the relative atomic displacements between top-layer and bottom-layer graphene as shown in Fig. 5.6c, d. In addition, we find that the strain distributions in both the suspended and supported regions by MD simulations can be well fitted to our analytical methods, validating our theoretical model for analyzing the experimental results.

5.3.3 Measurement of Interlayer Shear Stress

Based on the AFM and Raman measurements, the relationship between Δp and ρ is obtained in Fig. 5.7a. Clearly, the observed growth behavior of the interlayer shear zone is governed by an τ_2 of 0.02–0.06 MPa, two orders of magnitude lower than that of the graphene/substrate interface. Similar shear stresses have been reported for MWCNTs and graphite, demonstrating the ultralow interlayer shear interactions dominated by vdW forces [56, 57]. Compared to monolayer graphene bubbles, more significant shear deformation occurs between two graphene layers for bilayer graphene bubbles. Consequently, the bilayer graphene bubbles exhibited a markedly nonlinear relationship between $k_0\frac{\delta^3}{r_0^4}$ and Δp (Fig. 5.7b). Similar behaviors could also be observed in a previous work on few-layer graphene sheets [24]. Thus, the effects of the weak interlayer shear resistance should be considered in efforts to predict the overall bulging deformation of bilayer and few-layer graphene sheets.

Due to the relatively weak interactions between graphene layers, the common assumption that a multilayer graphene could be treated as a single sheet with an in-plane stiffness of nEt (n is the number of layers) may not be valid for the pressurized bubbles. Here, we developed a model for bilayer graphene bubbles based on the deformation analysis of individual graphene layers, taking into account the effect of

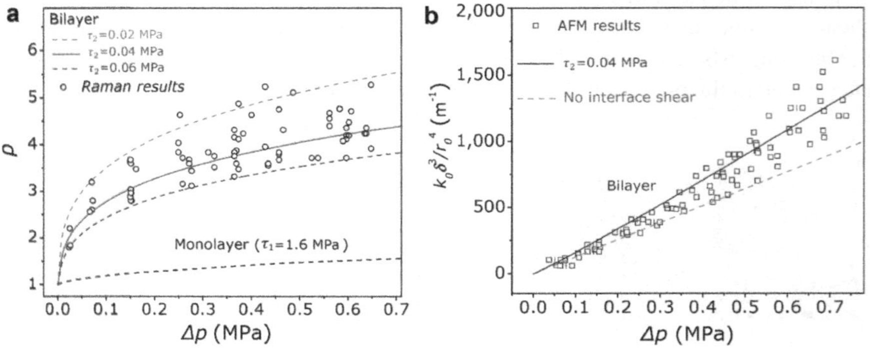

Fig. 5.7 **a** The radius ratio ρ of the interfacial shear zone versus the applied pressure difference for different values of interfacial shear stress. **b** The normalized center deflection $k_0\delta^3/r_0^4$ versus the applied pressure difference

a weak interlayer shear stress [2]. Let τ_2 be the interlayer shear stress in a bilayer graphene, while τ_1 is the interfacial shear stress between graphene and the substrate. The latter is assumed to be the same as the interfacial shear stress for the monolayer graphene. The measured shear zone radius for bilayer graphene was much larger than that for the monolayer, suggesting that $\tau_1 \gg \tau_2$. In such a case, a bilayer graphene bubble may be treated as superposition of two monolayer graphene bubbles, each subjected to a pressure difference in the suspended region and an interfacial shear stress in the supported region (Fig. 5.8). Within the hole, the interlayer shear stress between the two graphene layers is ignored, and the interactions between the two layers are predominantly in the normal direction, represented by a pressure difference (Δp_2) for the upper layer. The pressure difference for the lower graphene layer is then $\Delta p_1 = \Delta p - \Delta p_2$, where Δp is the applied total pressure difference. Meanwhile, the center deflections are assumed to be the same for both layers, as illustrated in Fig. 5.8.

With the interfacial shear stress (τ_1) and the in-plane stiffness (Et) determined from the monolayer graphene bubbles, we can determine the interlayer shear stress τ_2 using the measurements of shear zone radius (ρ) for the bilayer graphene bubbles as follows: (i) Determine $\zeta_2 = \bar{q}^{-\frac{2}{3}}/\bar{\tau}_2$ (Fig. 5.9a) for each ρ based on the master curve in Fig. 4.8b; (ii) Determine the coefficient k_2 (Fig. 5.9b) for each ζ_2 according to the master curve in Fig. 4.8c; (iii) Calculate the pressure difference Δp_2 for the

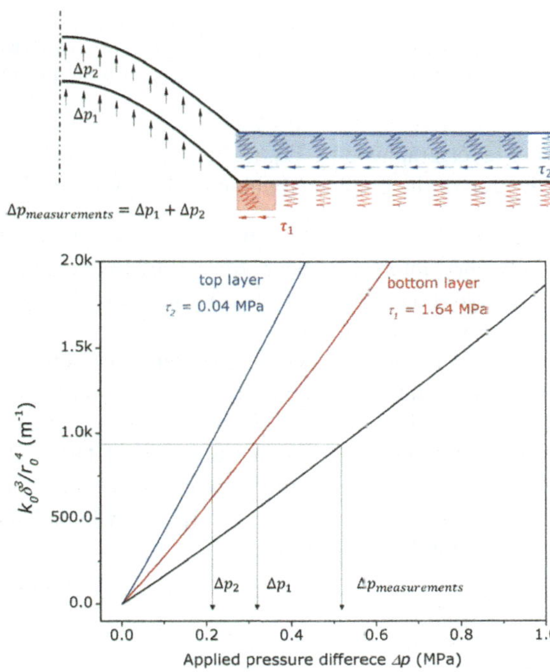

Fig. 5.8 Illustrations of a bilayer model via superposition of two monolayers, each following the membrane analysis but with different interfacial shear stresses outside the hole. For each center deflection, the total pressure difference equals the sum of the two pressures, i.e., $\Delta p(\delta) = \Delta p_1(\delta) + \Delta p_2(\delta)$

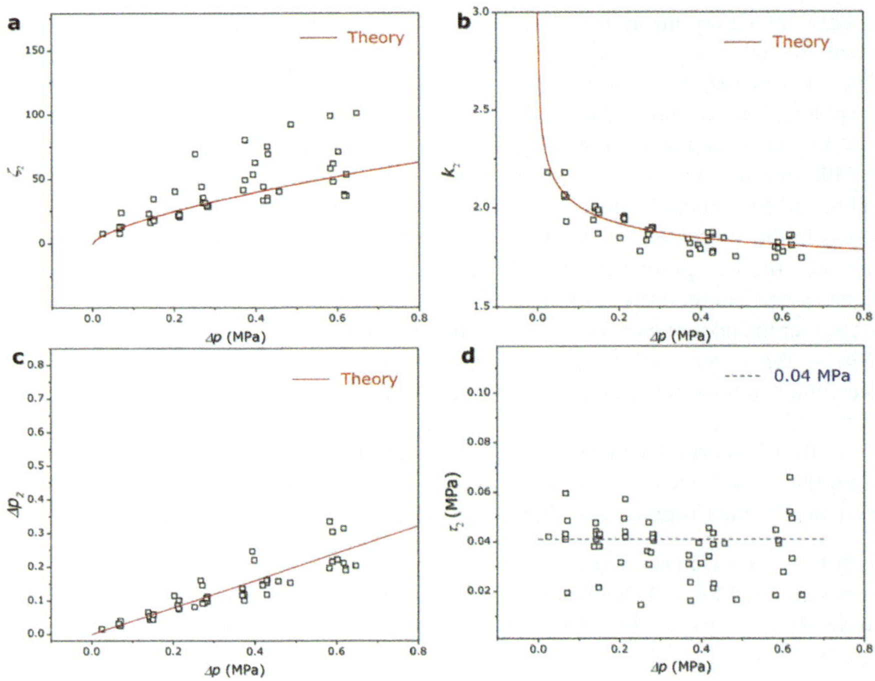

Fig. 5.9 The parameters extracted from monolayer measurements: **a** ζ_1 versus Δp, **b** k_1 versus Δp, **c** Et versus Δp, and **d** τ_1 versus Δp. The red lines in (**a**) and (**b**) are derived from the master curves in Fig. 4.14 using the average values of Et and τ_1 from (**c**) and (**d**), respectively

upper layer graphene by Eq. (5.11), i.e.,$\Delta p_2 = k_2 Et \frac{\delta^3}{r_0{}^4}$, using the measured center deflection of the graphene bubble as shown in Fig. 5.9c; (iv) Calculate the interlayer shear stress by $\bar{\tau}_2 = \bar{q}^{-\frac{2}{3}}/\zeta_2$ with $q = \Delta p_2$ and Et =390 N/m. As shown in Fig. 5.9d, the obtained interlayer shear stresses are averaged as 0.04 MPa. To the best of our knowledge, this is the first measurement of the interlayer shear stress between the atomically thin structures.

Moreover, for each measured center deflection, we can calculate Δp_1 using Eq. (4.14) with the interfacial shear stress τ_1 for the bottom graphene layer as shown in Fig. 5.9a. Then, the total pressure, $\Delta p_1 + \Delta p_2$, can be compared to the applied pressure Δp (Fig. 5.9b) as a validation of the analysis. With the obtained shear stresses, τ_1 and τ_2, the center deflection–pressure relation for the bilayer graphene can be predicted as $\Delta p = (k_1 + k_2) Et \frac{\delta^3}{r_0{}^4}$, which can be compared directly with the experimental data (Fig. 5.7b).

5.4 Interlayer Shear Effect on Bending of van der Waals Materials

5.4.1 *Breakdown of Continuum Mechanics*

Interlayer shear effect has been shown to be a dominant mechanism for stress relief at vdW interfaces, which is expected to have a substantial influence on the mechanical properties of multilayer 2D materials, such as Young's modulus, strength and mechanical resonance [31, 32, 58–60]. Here we focus on the bending rigidity, another crucial mechanical parameter governing the deformation of materials, which yet received much less attention compared with Young's modulus due to the difficulty in experimental characterization. In fact, the thinness of 2D materials makes them susceptible to out-of-plane deformations in response to external perturbations, forming buckles [61, 62], wrinkles [63, 64], crumples [65, 66], ripples [67, 68], scrolls [69, 70], folds [71, 72], bubbles [24, 36, 73–75], tents [76–78], and so on. To manipulate and control these out-of-plane deformations, a full understanding of the mechanics and underlying mechanisms of bending these atomically layered structures is required, which is actually dominated by the bending rigidity [79].

Classical plate theory assumed that an elastic plate resists bending by tension and compression on the opposite sides of a middle plane [80]. The bending rigidity of an elastic plate is related to its thickness (t) and the elastic moduli (E) of the material by $D = Et^3/12(1 - \nu^2)$, where ν is Poisson's ratio. This relation holds well for a perfectly glued multilayer with N identical layers such that the overall bending rigidity D is proportional to N^3. However, the classical relation is expected to break down for multilayer 2D materials for at least two reasons: (i) The mechanical resistance of the monolayer 2D material to bending and stretching deformation arises from different physical origins [81–84]. In graphene, for example, the in-plane Young's modulus is related to the in-plane σ bonding, while the bending resistance is attributed to the distortion of out-of-plane π bonds [81]. (ii) The vdW interactions between the atomic layers are weak in resisting interlayer shear/sliding [85–87], as demonstrated by recently observed self-rotation [88, 89] and self-sliding [27, 72] between 2D layers. The interlayer shear stiffness of bulk 2D materials (e.g. graphite) is typically orders of magnitude lower than their in-plane Young's modulus [90]. This extreme mechanical anisotropy may subject multilayer 2D materials to the interlayer sliding. In the limiting case when the interface is ultra-lubricated, each layer would bend independently such that the overall bending rigidity of the multilayer would scale linearly with the number of layers, $D \sim N$.

Despite extensive research into the mechanical properties of 2D materials for more than a decade, significant controversy on the bending rigidity of monolayer and multilayer 2D materials still exists [63, 81, 84, 91]. To reveal the effect of interlayer coupling on the bending rigidity of 2D materials, we performed pressurized bulging tests on three different 2D materials: graphene, hexagonal boron nitride (hBN), and molybdenum disulfide (MoS_2), as representatives of metallic, insulating and semiconducting 2D materials, respectively.

5.4.2 Membrane-to-Plate Transition

To achieve the bending rigidity of 2D materials, the measurements are enabled by controlling the bulged deflection of the 2D material to be comparable with its thickness, which can be realized by increasing the layer number of 2D materials up to 70 and reducing the cavity radius down to 0.5 μm. Figure 5.10 illustrates the pressurized bubble device that can create axisymmetric out-of-plane deformations of 2D materials. We first observe the shape characteristics of 2D material bubbles. Classical plate theory predicts the deflection of a plate responding to a uniform pressure as follows:

$$\frac{w(r)}{h} = \left(1 - \frac{r^2}{a^2}\right)^{\alpha} \tag{5.2}$$

where $w(r)$ is the out-of-plane deflection profile, h is the bubble height (deflection at the center), r is the distance from the hole center, a is the hole radius, and the power α is 2 for a linear elastic plate. However, such plate-like profile is rarely observed in 2D material bubbles because of the extreme thinness and flexibility of 2D materials [24, 36, 73–75]. Alternatively, the membrane limit is commonly adopted for these 2D material bubbles, where the bending rigidity is assumed to be negligible and the deflection profile of the bubble is approximately taken as a spherical cap, i.e., $\alpha = 1$ in Eq. (5.2).

We show in Fig. 5.10b experimentally measured profiles from graphene bubbles with a variety of thicknesses and heights. We normalized the deflection by the bubble height and the radial position by the bubble radius. For the bubbles with radii of 0.5 μm, the normalized deflection profiles collapse to one curve as predicted by the membrane analysis when the ratio between the bubble height and the thickness of the multilayer, h/t, is larger than 2—the case of most previous works [24, 36, 73–75]. For the bubbles with the height comparable to the thickness, i.e. $h/t \lesssim 1.5$, the bending effect becomes significant so that the deflection profiles collapse onto a different curve predicted by the plate theory. This observation implies that membrane-like behavior can transit to plate-like behavior for the bubbles of multilayer 2D materials

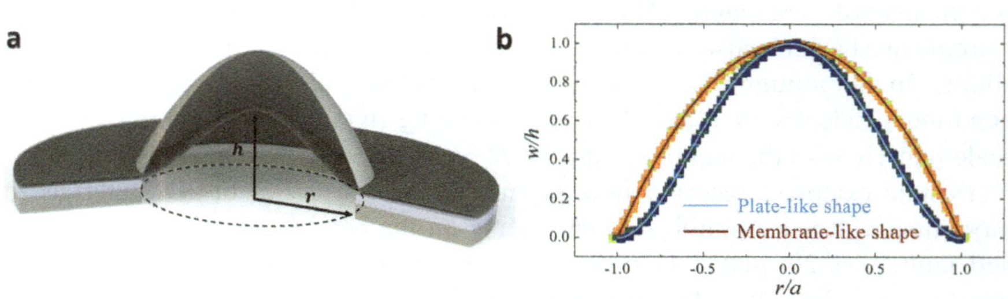

Fig. 5.10 **a** The schematic of pressurized bubbles with two distinct shape characteristics at small and large deflections. **b** Normalized deflection profiles measured from two sets of 0.5-μm-radius bubbles

by simply tuning the h/t. The membrane-plate transition also makes it possible to simultaneously determine the in-plane Young's modulus and bending rigidity of the 2D multilayers.

We then consider the pressure-height relation for these multilayer 2D material bubbles. Previous numerical studies proposed a nonlinear plate analysis for pressurized bubbles that included both in-plane stretching and bending of the 2D material, leading to [92, 93]

$$\frac{\Delta p a^4}{h^3 t} = A(\nu)E + \frac{64D}{t^3}\left(\frac{t}{h}\right)^2 \tag{5.3}$$

where $A(\nu) \simeq \left(0.7179 - 0.1706\nu - 0.1495\nu^2\right)^{-3}$, which is 3.10 for graphene with $\nu = 0.165$, 3.28 for hBN with $\nu = 0.221$ [94], and 3.46 for MoS_2 with $\nu = 0.27$ [22]. Note that Eq. (5.3) was found to be a good approximation even for monolayer graphene bubbles by using the 2D in-plane stiffness ($E_{2D} = Et$) and a bending rigidity (D) that is independent on E [92]. We also note that the interfacial shear deformation between the supported 2D material and the SiO_2 substrate is negligible in our experiments because of the relatively small pressure and deflection, and hence the clamped boundary condition at the edge of the bubble is reasonable. In Fig. 5.11, the dot-dashed lines represent the membrane limit of Eq. (5.3) where the bending term is neglected. The solid lines represent the idealized nonlinear plate model assuming a bending rigidity $D = Et^3/12(1-\nu^2)$ such that the response ($\frac{\Delta p a^4}{h^3 t}$ vs. h/t) does not depend on the thickness. Evidently, the experimentally measured responses for all three 2D materials approach the membrane limit when the ratio h/t is greater than 2. For $h/t < 1.5$, the bubble responses exhibit a transition to the plate-like behavior, but deviate considerably from the idealized nonlinear plate solution. This observation implies that the use of the classical $D - E$ relation would overestimate the bending rigidity of the multilayer 2D materials.

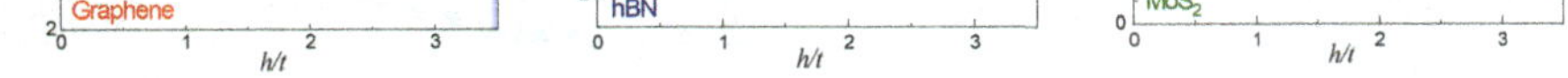

Fig. 5.11 $\Delta pa^4/h^3t$ as a function of h/t for graphene **a**, hBN **b** and MoS_2 **c**, respectively. The solid lines represent the idealized nonlinear plate solution, and horizontal dot-dashed lines are the membrane solution. In the color system, darker symbol denotes the thicker sample

5.4.3 Measurement of Elastic Properties

Equation (5.3) enables us to determine the Young's modulus and bending rigidity of the multilayer 2D material as well as their dependence on the layer thickness. We start with Young's moduli of multilayer 2D materials by focusing samples under relatively large pressure [93]. In this case, h/t is relatively large and the second term on the right-hand side of Eq. (5.3) is negligible so that Young's modulus can be determined directly by measuring $\frac{\Delta pa^4}{h^3t}$. In Fig. 5.11, for graphene, hBN, and MoS_2 bubbles with $h/t \gtrsim 2$, the data for $\frac{\Delta pa^4}{h^3t}$ collapse onto the dot-dashed lines, as predicted by the membrane limit. The domain of h/t for the membrane-like behavior is consistent with our shape measurements in Fig. 5.10. We then extract Young's moduli of the multilayer 2D materials as summarized in Fig. 5.12a. We find that the average values are 939.5 ± 21.7 GPa, 769.9 ± 12.9 GPa, and 314.3 ± 8.4 GPa for the multilayer graphene, hBN, and MoS_2 with the number of layers ranging from ~10 to ~70. For each 2D material, the extracted Young's modulus shows minor dependency on the thickness, and the average values agree well with those measured by AFM indentation of monolayers [22, 48, 95].

When the height of deformed bubble becomes comparable to the thickness of 2D multilayers, contribution from bending substantiates and both terms on the right-hand side of Eq. (5.3) should be accounted for. A membrane-to-plate transition is expected to occur in the mechanical response of bubbles (Fig. 5.10). With the thickness-independent Young's modulus extracted for the 2D materials from the membrane model, we can then calculate the bending rigidity D by comparing Eq. (5.3) with the measured response $\frac{\Delta pa^4}{h^3t}$ for bubbles with relatively small deflection ($h/t \lesssim 1.5$). Figure 5.12b summarizes the values of D for graphene, hBN, and MoS_2 multilayers. As expected, the bending rigidity increases with the thickness of 2D materials. However, the thickness dependence does not follow either one of

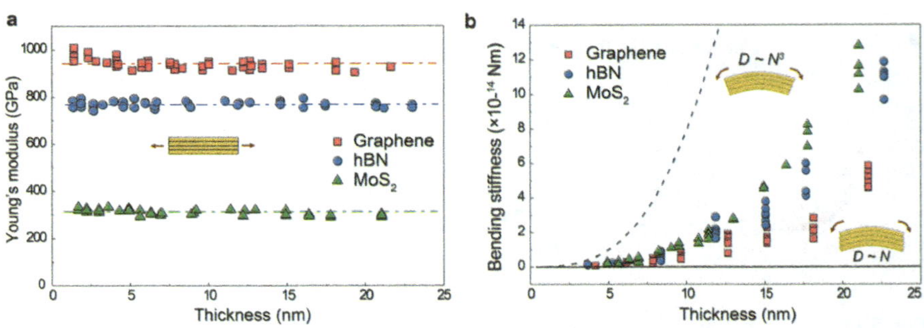

Fig. 5.12 Extracted Young's modulus (**a**) and bending rigidity (**b**) as a function of thickness. At a given thickness, AFM measurements of up to 17 different bubbles produce the results plotted as symbols. The dash-dotted lines in (**a**) indicate the average values, while the dashed and solid lines in (**b**) are the theoretical predictions for the perfectly glued ($D \sim N^3$) and ultra-lubricated ($D \sim N$) cases, respectively

the two limiting cases. Specifically, the bending rigidity of a multilayer graphene ($t > 5$ nm) is much lower than that expected from classical plate theory ($D \sim t^3$; dashed line in Fig. 5.12b), but much higher than the ideally lubricated cases ($D \sim t$; solid line in Fig. 5.12b). More interestingly, for samples with similar thickness, we find $D_{\mathrm{MoS2}} > D_{\mathrm{hBN}} > D_{\mathrm{graphene}}$, whereas the values of Young's moduli are ordered inversely, that is, $E_{\mathrm{graphene}} > E_{\mathrm{hBN}} > E_{\mathrm{MoS2}}$. This again indicates that the bending rigidity of multilayer 2D materials violates the predictions of the classical plate theory.

5.4.4 Quantification of Interlayer Shear Effect

We attribute the abnormal bending behaviors of multilayer 2D materials to the effect of interlayer shear/slippage between atomic sheets, which breaks down the assumptions in classical plate theory. The interlayer sliding of bent 2D material multilayers has been reported by both simulations and experiments in literature [96, 97]. While this mechanism may be qualitatively simple to understand, the quantitative measurement of its effect on the bending properties of a 2D material multilayer has been challenging and has not been reported to our best knowledge. To shed further light on such effects, we propose a modified formula for the effective bending rigidity of multilayer 2D materials as

$$D_{\mathrm{eff}} = f(N) \cdot \frac{Et^3}{12\left(1 - \nu^2\right)} \tag{5.4}$$

where $f(N)$ is a function of the layer number and depends on the strength of interlayer shear interactions, as a result of competition between the intralayer stretching/compressing and the interlayer shear/sliding. Theoretically, there are two limiting cases. The upper bound assumes perfectly bonding so that the interlayer slippage is prohibited (Fig. 5.13a) while the lower bound corresponds to frictionless interlayer interactions or ultra-lubricated (Fig. 5.13b). In case that the perfect

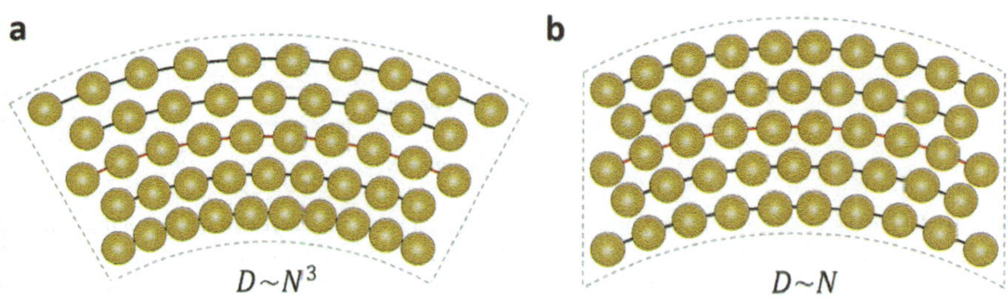

Fig. 5.13 The microstructural deformation upon bending of multilayer 2D materials with perfectly glued (**a**) and ultra-lubricated (**b**) interfaces

bonding between layers is assumed, the bending rigidity of multilayer 2D materials is [98]

$$D_{\text{eff}} = ND_1 + \frac{Et_1{}^3}{12(1-\nu^2)}(N^3 - N) \tag{5.5}$$

where D_1 is the bending rigidity of a monolayer, with values of 1.60 eV, 1.29 eV and 9.9 eV reported for graphene, hBN and MoS_2, respectively [71, 81, 99, 100]. Based on Eq. (5.5), we obtain

$$f(N) = \frac{12D_{\text{eff}}(1-\nu^2)}{Et_{\text{total}}^3} = 1 - \frac{1}{N^2} + \frac{12(1-\nu^2)D_1}{Et_1{}^3N^2} \tag{5.6}$$

where $t_{\text{total}} = Nt_1$ and t_1 is the interlayer distance. Alternatively, for perfectly lubricated layers, the bending rigidity is simply

$$D_{\text{eff}} = ND_1 \tag{5.7}$$

and thus

$$f(N) = \frac{12(1-\nu^2)D_1}{Et_1{}^3N^2} \tag{5.8}$$

The theoretical limits are plotted as solid lines for graphene in Fig. 5.14. Experimentally, the function $f(N)$ can be determined based on Eq. (5.3):

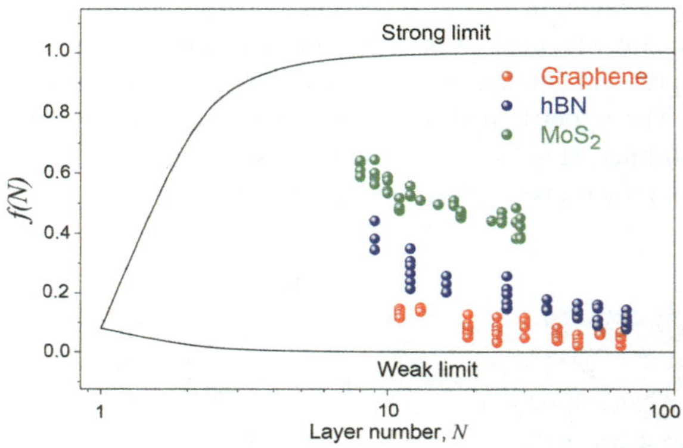

Fig. 5.14 The layer-dependent function to measure the effect of interlayer coupling on the bending rigidity of multilayered graphene, hBN, and MoS_2. The black lines are theoretical predictions of the limiting cases for graphene. Our experimental results are denoted by colored markers, sitting between the two limits

$$f(N) = \frac{3(1-v^2)h^2}{16t^2}\left(\frac{\Delta pa^4}{Eth^3} - A(\nu)\right) \tag{5.9}$$

as shown in Fig. 5.14 for multilayer graphene, hBN and MoS_2. Evidently, the values of $f(N)$ for these multilayer 2D materials lie between the two theoretical limits. These results imply that both intralayer stretching/compressing and the interlayer shear/sliding happen when bending the multilayer 2D materials. Among the three 2D materials, the graphene multilayers have the lowest values of $f(N)$ and the MoS_2 multilayers have the highest value. Despite the highest in-plane elastic modulus of graphene, the bending rigidity of multilayer graphene is the lowest due to the interlayer slippage, close to the lower limit. It is expected that the modified formula for the bending rigidity in Eq. (5.4) not only works very well for all our samples, but also applies, more broadly, for other multilayer vdW materials and their homo/heterostructures. The methods and data provided in this work offer quantitative measures for the elastic responses of multilayer 2D materials through a simple function $f(N)$. This function may quantify the level of interlayer slippage, as well as the coupling between stretching and bending, which highly depend on the types of the 2D material and its multilayer structures.

5.5 Tuning Interlayer Coupling by Boron Doping

5.5.1 Structural Characterization of Boron Doped Graphene

Chemical doping via incorporating heteroatoms into the crystal lattice tends to be an effective way to tailor interlayer coupling of multilayer graphene systems. Here we conduct the first mechanical investigation into the effects of boron doping on the interlayer coupling behaviors of multilayer graphene system. We use natural graphite as a host material for boron doping following previously reported strategies [47]. X-ray photoelectron spectroscopy (XPS) was utilized to verify the presence of the boron atoms and identify the boron content in graphite, ranging from 0.60 to 3.09% (Fig. 5.15a). As shown in Fig. 5.15b, the C1s peak located at 284.8 eV

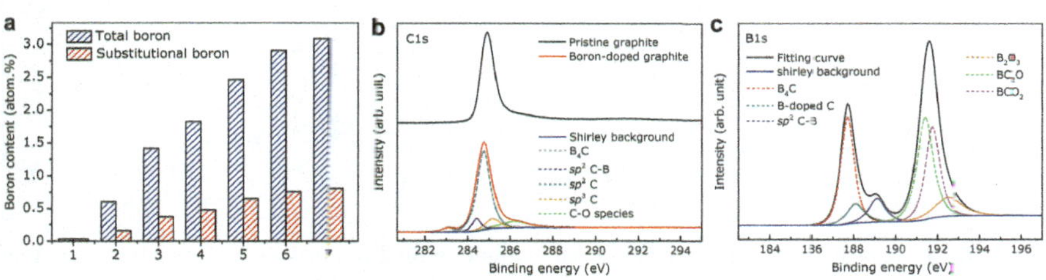

Fig. 5.15 **a** Seven prepared samples with different boron content in graphite. **b** The C1s XPS spectra of pristine and boron-doped graphite. **c** The B1s XPS spectra of boron-doped graphite

was attributed to the sp^2 carbon atoms and there appeared a slight downshift to 284.6 eV after boron doping, which corresponds to the lowering Fermi level through the formation of C–B bonds [47]. Besides, a new peak arose at 283.2 eV originates from the nonstoichiometric boron carbide (B_xC_y), especially B_4C, according to previous work [101]. Two distinct features at 284.4 and 285.3 eV are assigned to substitutional boron atoms into the honeycomb lattice and sp^3-carbon atoms, respectively. In addition, the component at 286.2 eV may correspond to the C–O species resulted from the contamination during thermal annealing [101]. The strong peak in the B1s spectral region also corroborates the successful boron doping of the graphite as shown in Fig. 5.15c, in which the components centered at 187.7 and 188.1 eV are attributed to B_4C and boron-doped carbon nanoparticles, respectively. The peak at 189.1 eV can be associated to the sp^2 C–B bonds present in graphene, indicative of the boron substitution [102]. According to the curve fitted areas for B1s spectra, the solubility of substitutional boron could hence be deduced, with the maximum of ~0.81%, and the *x*-coordinate "boron content" in following figures actually indicates the amount of substitutional boron atoms in graphene. At higher binding energy than 190 eV, three fitting peaks have been identified and denoted as different boron-oxygen-carbon compounds (C_2–BO at 191.4 eV and C–BO_2 at 191.8 eV) and boron oxide (B_2O_3 at 192.5 eV). Note that, as shown in Figure S1b, the oxygen species induced characteristic peak at ~191 eV can be only visible in the samples with boron content higher than 0.76%, which possibly results from with overweight ratio of boron oxide to graphite and longer reaction time. As a matter of fact, the centroids of both C1s and B1s photoemission lines are slightly different from the reported results, wherein the discrepancy lies in distinct raw materials (e.g. natural graphite, graphene oxide, etc.), preparation methods (e.g. thermal annealing, chemical vapor deposition, electrothermal reactions, etc.) and consequently complex structures possessing oxy-boron-carbide species with different densities and distributions [103–109].

Figure 5.16a displays the results of XRD of boron-doped graphite, where the (002) diffraction pattern had two distinct diffraction peaks initially. These two peaks arose from the two different structures within the same graphite, including well-ordered structure and turbostratic structure [110]. With increasing boron content, the turbostratic fraction would be enhanced so that the two peaks tended to merge into

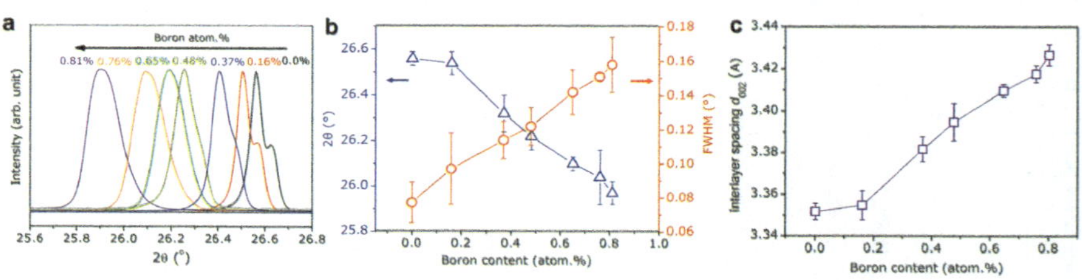

Fig. 5.16 **a** X-ray diffraction pattern of graphite samples with different substitutional boron content. **b** The diffraction angle and the full width at half-maximum (FWHM) as a function of the substitutional boron content. **c** The interlayer spacing of graphite as a function of the substitutional boron content

one peak with increasing FWHM up to 0.2°. Besides, the diffraction angle tends to decrease obviously as the boron content was improved (Fig. 5.16b), corresponding to an increase in the interlayer spacing between graphene planes, as depicted in Fig. 5.16c. This is in contrast to the previous results that the boron-doped graphite by substitution reaction shows a higher diffraction angle and shorter interlayer distance than that of pristine graphite [47]. As mentioned above, the variation of the interlayer spacing greatly depends on the competition between the weakened π-electron density and Poisson contraction effect, where the latter is responsible for the decrease in interlayer spacing [111, 112]. In our case, however, boron doping gives rise to disordered structures due to the random existence of oxy-boron-carbide species, so that the weakened π-electron density between graphene planes possibly plays a more significant role than Poisson contraction to induce the enlarged interlayer spacing. Similar results have been reported in other works on carbon nanotubes (CNTs), where the average interlayer spacing of the boron-doped double-walled CNTs is enhanced to be 0.35–0.41 nm [113]. Additionally, Raman measurements were also performed to evidence the effective doping in graphite, where the stiffening of the Raman D-band and the degradation of the Raman 2D-band can be clearly observed in Fig. 5.17.

Boron-doped graphene flakes were exfoliated by micromechanical cleavage of boron-doped graphite. Figure 5.18a, b show the low- and high-resolution transmission electron microscopy (TEM) images of boron-doped bilayer graphene, displaying no distinctive differences in the texture compared with those of pristine ones [114]. The crystallinity is identified by the selected area electron diffraction (SAED) pattern in Fig. 5.18c, where the six-fold symmetry is clearly visible, suggesting hexagonal lattice structure of graphene is maintained after boron doping. Based on the measurement of distances between reflex spots, the atomic separation in boron-doped graphene can be deduced as 1.59 Å. This value is relatively larger than the original

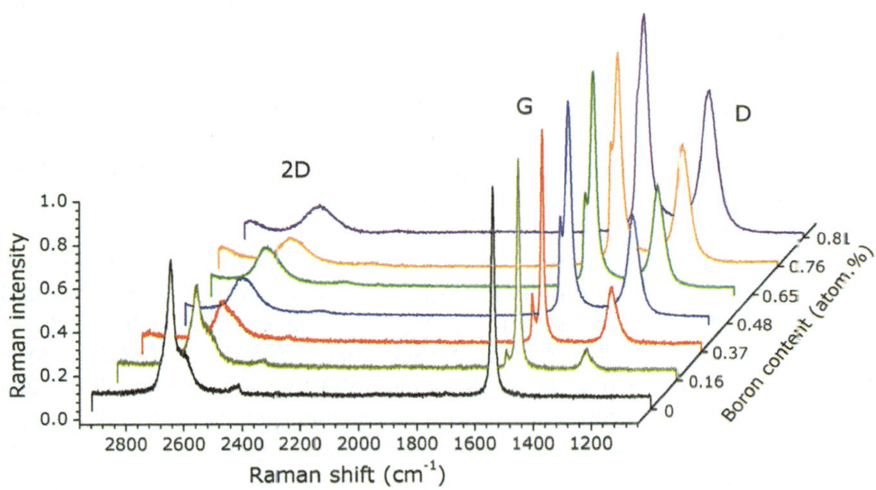

Fig. 5.17 Raman spectra of graphite samples with different boron content

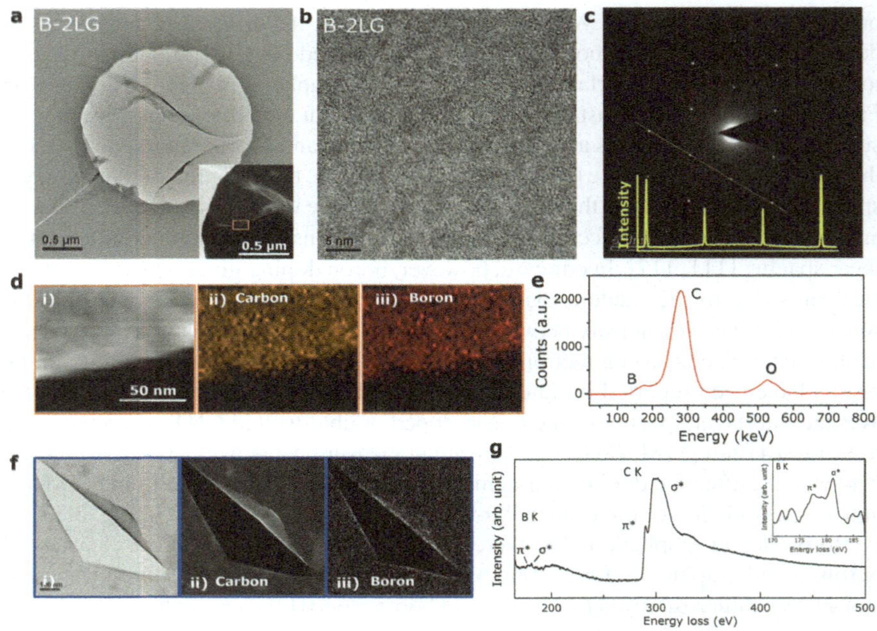

Fig. 5.18 **a** TEM image of boron-doped bilayer graphene (B-2LG) with substitutional boron content of ~0.65%. The inset is the zoom-in HAADF image. **b** Atomic-scale TEM image of boron-doped bilayer graphene. **c** The SAED pattern and corresponding profile plot of diffraction peak intensities along the dashed line. (**d**-i) Dark field TEM image and EDS mapping of carbon (**d**-ii) and boron (**d**-iii) taken from the zoom-in region as marked by the rectangular in the inset of (**a**). **e** The corresponding average EDS spectrum obtained from the boron-doped bilayer graphene in (**d**-i). **f**-i Zero-loss image showing the confined crack in graphene and corresponding EELS mapping of **f**-ii carbon and **f**-iii boron from the area of (**f**-i). **g** K-shell excitations of boron and carbon are revealed from the core EELS spectra

bond length between carbon atoms (1.42 Å), which agrees well with ever reported results extracted from scanning tunneling microscopy (STM) [115]. Importantly, the line profile of the diffraction pattern shows the diffraction intensity ratio of the outer peaks over the inner peaks close to 2.0, indicating the retained Bernal-stacked structure as well [116, 117].

To further confirm the presence of boron element in boron-doped bilayer graphene, energy-dispersive spectroscopy (EDS) mapping was performed over the localized region in Fig. 5.18d-i, as marked in the high angle annular dark field (HAADF) image (Fig. 5.18a inset). Carbon (Fig. 5.18d-ii) and boron (Fig. 5.18d-iii) signals were detected uniformly over the whole region, validating our structural design and indicating of a homogeneous boron doping in the graphene plane. In contrast with carbon element, the seemingly analogous distribution of boron element is essentially attributed to strong background signals together with sparse boron atoms therein. As presented in the EDS spectrum (Fig. 5.18e), the carbon peak is greatly stronger than the boron peak, which implies the considerably low boron content of doped graphene flakes. The electron energy loss spectroscopy (EELS) measurement was also performed to validate the substitutional boron atoms doped in graphene with retained hexagonal structure. In detail, the spatially resolved EELS mappings are made as presented in Fig. 5.18f, where the carbon and boron elements coexist in the same regions, indicating of a relatively homogeneous distribution of boron atoms in graphene without any segregation domains. Figure 5.18g presents a typical EELS spectrum with two distinct features starting at ~180 and ~290 eV, corresponding to the characteristic K-shell ionization edges of boron and carbon, respectively [118]. The 1 s−π^* (290 eV) and 1 s−σ^* (~300 eV) peaks of the C K-edge are characteristic of sp^2-hybridrized carbon networks, agreeing well with the previous reports for graphene or carbon nanotubes [119]. An inspection of the B K-edge also shows discernible π^* and σ^* peaks, suggesting that the boron atoms are in the similar sp^2-hybridized state as their carbon counterparts [120]. This further proves that most of the boron atoms are substitutionally doped in graphene to retain its hexagonal structure.

The effect of boron doping on the surface structure is further visualized by scanning tunneling microscopy (STM). Obviously, the STM image of boron-doped multilayer graphene in Fig. 5.19b exhibits randomly dispersed bright areas (diameters ranging from 0.4 to 1.0 nm) aligned with previous results for boron-doped highly oriented pyrolytic graphite (HOPG) [115]. Such inhomogeneous distribution is never seen on pristine graphene surface, which displays a clear spot array with relatively uniform brightness (Fig. 5.19a). It must be assumed that the mottled patterns are stemmed from the substitution of boron atoms in the carbon planes, inducing the local distortions in both the topographic and electronic structures of graphene [47].

As a powerful non-destructive technique for identifying the doping of graphene [121], Raman spectra further provides additional evidence of the introduction of boron atoms. Figure 5.20a plots the Raman spectrum of boron-doped bilayer

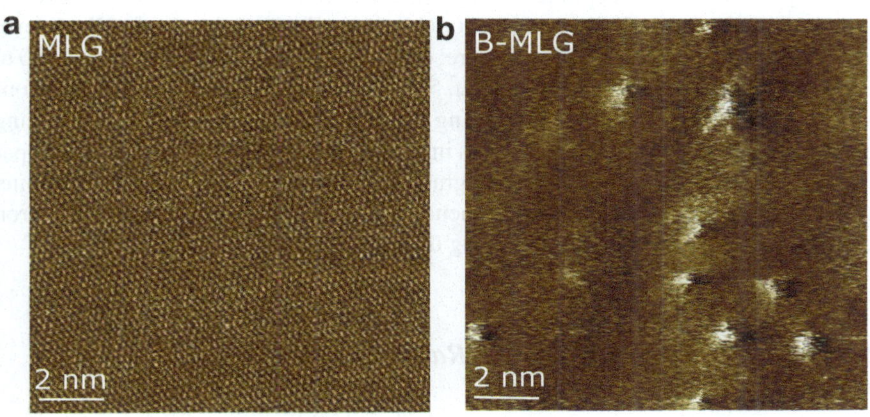

Fig. 5.19 STM images of **a** pristine and **b** boron-doped HOPG

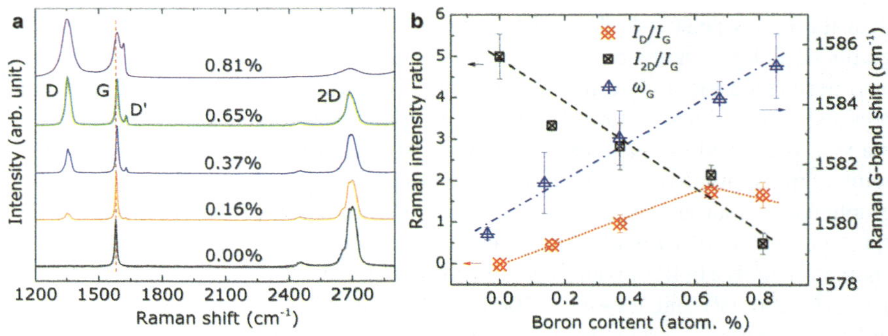

Fig. 5.20 **a** Raman spectra of bilayer graphene with various boron content. **b** Variations of Raman I_D/I_G ratio of bilayer graphene as a function of substitutional boron content

graphene at different values of boron content. Compared with pristine graphene merely having G-band and 2D-band in general, two new featured peaks are visible, named D-band and D′-band, which are bound up with Raman double resonance process induced by the substitution of boron atoms as defect sites in graphene structure [122]. As a result, both the Raman D-band and D′-band are enhanced simultaneously with increasing boron content. Remarkably, a plateau of I_D/I_G ratio can be observed at substitutional boron content >0.65% as shown in Fig. 5.20b, implying the saturation of substitutional boron doping as well as the formation of structural voids [47, 123]. This is consistent with the three-stage classification of disorder proposed by Ferrari and Robertson [124, 125], where the Raman intensity ratio I_D/I_G is proportional to the defect concentration at low defect density and inversely proportional at high defect density. On the other hand, boron doping as the hole doping, would change the Femi energy of graphene [126]. The consequently altered electron–electron scattering tends to cause the drastic variation of 2D-band intensity while G-band intensity is almost insusceptible [127, 128], corresponding to a remarkable decline ~90% of I_{2D}/I_G ratio relative to that prior to doping. Besides, the varying Femi level also leads to the upshifts in G-band. Here, only slight G-band upshifts from ~1580 to ~1585 cm^{-1} are visible, as shown in Fig. 5.20b. Such phenomenon was also demonstrated in Kim's work and the underlying mechanism possibly lies in the softening caused by boron-induced tensile strains in graphene lattice [129]. Given the comparable Raman features of boron-doped graphene exfoliated from the same graphite, the doping level of graphene flakes is hence represented by the substitutional boron content of host graphite in the following discussions.

5.5.2 *Interlayer Coupling via Raman Shear Mode*

Earlier works have investigated the in-plane performances of boron atoms incorporated graphene crystals [130–132], while the impact on the interlayer coupling from mechanical viewpoint has been unexplored yet. Recently, the low-frequency Raman spectroscopy has been uncovered to be a diagnostic tool to explore the interlayer coupling [20, 133–136]. For instance, in the case of graphite, there is merely one low-frequency E_{2g} mode denoted as C-band, which is responsible for the relative motion of atoms in two adjacent planes [137]. Tan et al.have revealed the Raman signature of the shear mode (C-band) of multilayer graphene and successfully measured the interlayer coupling strength as well as the interlayer shear modulus [22]. Furthermore, similar low-frequency Raman modes are observed in other layered materials, implying the universality of this technique can be expected for all 2D nanomaterials, such as MoS_2, WSe_2, and black phosphorus [135, 136, 138]. Thus, we employ Raman shear mode to provide a direct probe of interlayer interaction of boron-doped multilayer graphene.

Graphene flakes in different layer numbers were mechanical exfoliated on silicon wafer, as shown in Fig. 5.21a. Aligned with discussions in Sect. 5.5.1, boron doping introduces D- and D′-band in the Raman spectra while degrade the 2D-band simultaneously, as seen in Fig. 5.21c. Furthermore, for multiple layers, since the wave function consists of the wave functions for various layers, the scattered amplitude

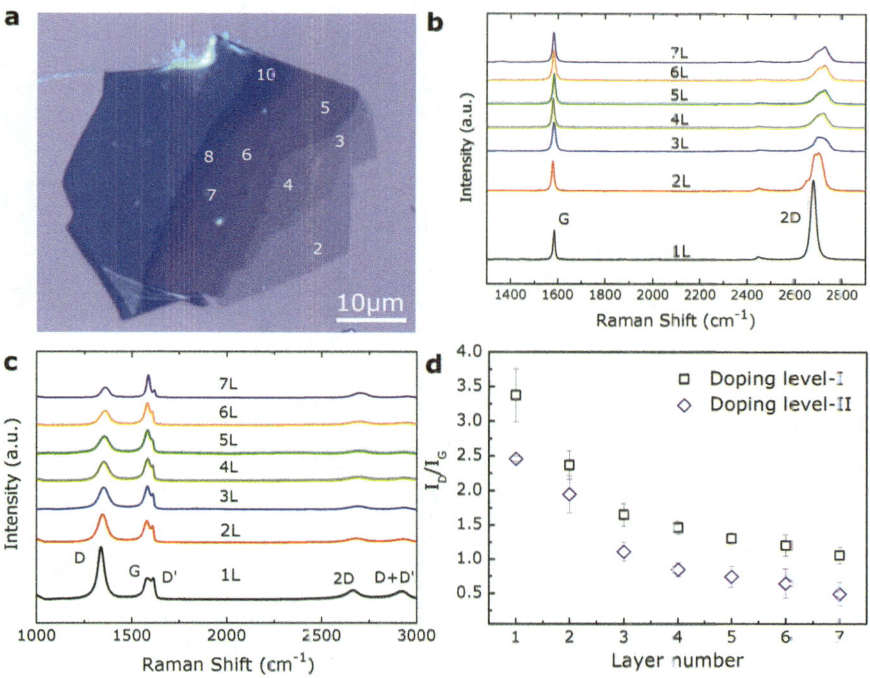

Fig. 5.21 **a** Optical image of the exfoliated graphene flakes with different layer numbers. Raman spectra of graphene flakes with different layer numbers from one single piece of **b** pristine graphite and **c** boron-doped graphite with substitutional boron content of ~0.81%. **d** Raman intensity ratio (I_D/I_G) as a function of layer numbers for graphene with different doping levels

becomes inversely proportional to the number of layers. Specifically, the scattering amplitude of a photoexcited electron by a boron atom in bilayer graphene is expected to be a half of the value for single graphene [47]. Thus, a decreasing I_D/I_G is expected as the layer number increases, as shown in Fig. 5.21d.

Figure 5.22a plots the low-frequency Raman spectra focusing on C-peak spectral region for a set of boron-doped graphene samples with increasing layer numbers. Apparently, the boron doping enables peak profiles to be fitted well with Lorentz functions, rather than the Breit-Wagner-Fano lineshape for pristine samples, suggesting the possible disappearance of the quantum interference between the C mode and a continuum of electronic transitions near the K point [20]. Herein, we use the notation NLG to indicate graphene with N layers (e.g. 1LG = monolayer). As the thickness decrease from 10 to 2LG, the C-peak tends to downshift from ~42 to ~29 cm^{-1}, as guided by dashed lines. Such variation of peak positions can be interpreted by the

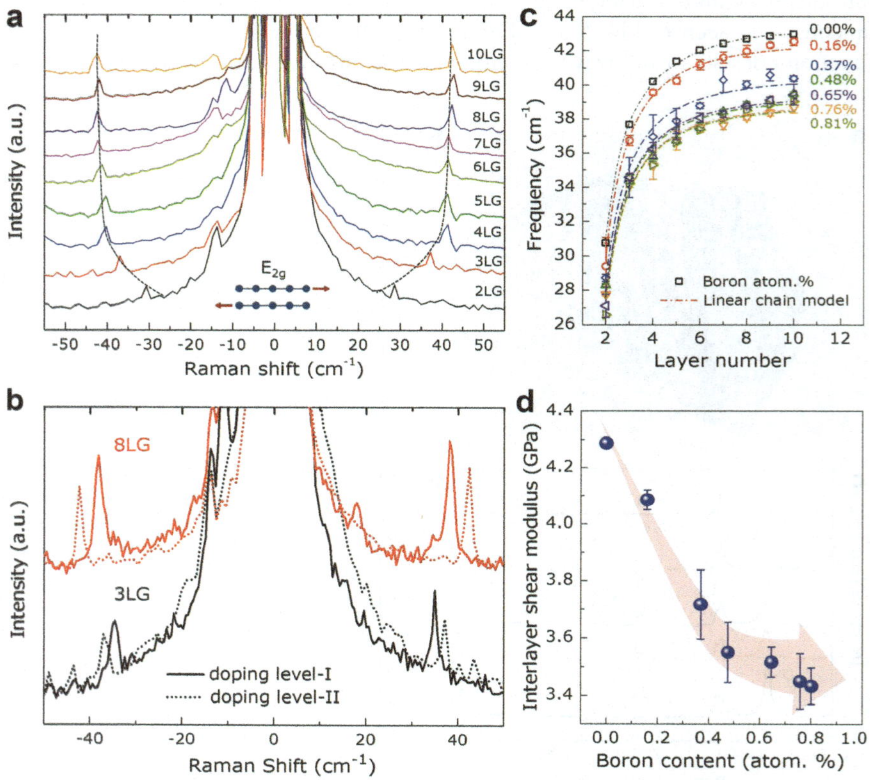

Fig. 5.22 **a** Stokes and anti-Stokes Raman spectra for the C-peak spectral region with increasing graphene layer numbers (substitutional boron content ~0.16%). The inset is the vibrational normal mode of interlayer shear mode in multilayer graphene. **b** Peak positions of C band as a function of layer numbers for samples with different boron content. **c, d** The calculated interlayer shear modulus as a function of boron content

linear chain model, essentially corresponding to the interlayer interactions of multilayer graphene [20]. Note that, it has been demonstrated that the boron doping, classified as p-doping, typically results in upshift in Raman peaks as well [139]. However, herein, doping effect induced C-band shifts could be negligible as the consequent Raman shift is almost proportional to the corresponding peak positions. We can assert the reliability of utilizing Raman C-band shifts to elucidate the interlayer shear behaviors.

Typically, the dependence of Raman C-band position on the number of layers can be explained by considering a linear-chain model [20, 140, 141]. For multilayer graphene with N layers, there are $2N$ atoms per unit cell. The corresponding in-plane optical modes consist of N degenerate pairs of in-plane stretching modes, and $(N-1)$ degenerate pairs of in-plane shear modes between neighboring layers. Considering a simplified case, we assume that one layer interacts strongly only with adjacent layers and that the strength of this interlayer coupling is characterized by an interlayer force constant per unit area (α). According to the theory of lattice dynamics, the frequency ω_i of the ith vibrational mode can be expressed as [142]:

$$(\omega_i)^2\mathbf{u_i} = \frac{1}{2\pi^2c^2\mu}\mathbf{D}\mathbf{u_i} \tag{5.10}$$

where $\mathbf{u}_i$ is the eigenvector of the ith vibrational mode, μ =7.6 × 10^{-27} kg Å^{-2} is the mass per unit area for monolayer graphene, $c = 3.0 \times 10^{10}$ cm/s is the speed of light, $\mathbf{D}$ is the force constant matrix of the shear mode. The solution to Eq. (5.10) is

$$\omega^2_{C,NN-i} = \frac{\alpha}{2\pi^2c^2\mu}\left\{1 - \cos\left[\frac{(i-1)\pi}{N}\right]\right\} \tag{5.11}$$

where $i = 1, 2 \ldots N - 1$. The corresponding ith displacement eigenvector is given by

$$v_j^{(i)} = \cos\left[\frac{(i-1)(2j-1)\pi}{N}\right] \tag{5.12}$$

where j labels the layers. By setting $i = N$, the peak position of C-band for a given layer number is obtained by

$$\omega(C) = \frac{1}{\sqrt{2}\pi c}\sqrt{\frac{\alpha}{\mu}}\sqrt{1+\cos\left(\frac{\pi}{N}\right)} \tag{5.13}$$

We repeated detecting the Raman shear modes of boron-doped graphene samples with different boron contents and extracted the Raman C-peak position using Lorentz fit, as shown in Fig. 5.22b. In addition to the thickness dependence as expected, a declined tendency of C-peak position is observed with increasing doping level. By fitting the experimental data (dash-dotted line in Fig. 5.22b), the obtained interlayer force constant which decreases from $12.8 \times 10^{28}\mathrm{N/m^3}$ to $10.2 \times 10^{28}\mathrm{N/m^3}$ with increasing boron content. In turn, such a weakening of interlayer interactions would modify the electronic dispersion, leading to the resonant enhancement of C-band intensity (solid lines in Fig. 5.22c) [143].

Considering a uniform distribution of shear force (F), it can be deduced from

$$F = \alpha A x \tag{5.14}$$

where A is the graphene surface area and x is the shear displacement. From the perspective of elastic mechanics, the shear stress (τ) and shear strain (γ) under small deformation can be expressed as:

$$\tau = F/A \tag{5.15}$$

$$\gamma = x/t \tag{5.16}$$

where t is the equilibrium distance between two adjacent graphene layers. Combining Eqs. (5.14–5.16), we have the interlayer shear modulus

$$G = \tau/\gamma = \alpha t \tag{5.17}$$

As presented in Fig. 5.22d, the interlayer shear modulus drops by about 20% compared with the initial value at boron content lower than 0.65% and remains nearly constant with further increasing concentration. Such a two-stage feature can be actually interpreted by the structural evolution as manifested in XPS spectrum. It is envisioned that, at boron content from 0.16 to 0.65%, the substitutional boron plays the dominant role in the weakening of the interlayer coupling. As the boron content increases up to 0.76%, the saturation of substitutional boron solubility is nearly reached while additional surface species with increasing oxygen concentration have negligible influence on the interlayer coupling, so that a sluggish decreased tendency is expected as shown in Fig. 5.22d. Besides that, the absolute values are consistent with results of previous works [144, 145], implying the perfect Bernal-stacking of graphene flakes in our work. The reduced interlayer shear modulus is also in agreement with the enlarged interlayer spacing of graphite from the XRD results in Fig. 5.16.

A better insight into the underlying mechanism in above-discussed experimental results can be obtained by performing MD simulations to clarify the interlayer behavior of multilayer graphene after boron doping. The schematics of boron-doped multilayer graphene system are presented in Fig. 5.23a, where boron atoms are randomly distributed in each layer. Expectedly, the substitutional boron atoms can weaken the π-electron density and hence the interactions between graphene layers, enlarging the interlayer spacing. Consequently, the interlayer spacing (d_{002}) exhibits an increasing tendency with the growing boron content as seen in Fig. 5.23b. For instance, d_{002} for pristine graphene is 3.418 ± 0.006 Å while it is enhanced up to 3.438 ± 0.007 Å for graphene with boron content of ~3.0%. Note that, in our experiments,

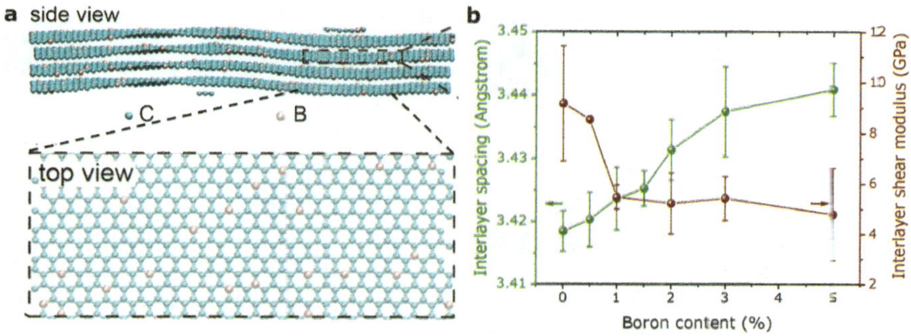

Fig. 5.23 **a** Schematics of boron-doped multilayer graphene with side view and top view. **b** Normalized interlayer spacing and interlayer shear modulus as a function of the boron content

d_{002} increases from 3.352 ± 0.004 Å for pristine graphite to 3.427 ± 0.005 Å once reaching boron content of 3.0%. Such difference is definitely acceptable because in simulations, boron atoms are entirely substitutional-type in graphene while it could not be guaranteed in experiments as some isolated atoms or clusters might be produced during thermally treating process, as evidenced in Fig. 5.15a. But it should be emphasized that, it is the substitutional boron rather than the out-of-plane species that takes the main responsibility for the enlarged interlayer spacing. In efforts to compare with Raman results, we also applied shear deformations into the system to directly reveal interlayer shear modulus. In Fig. 5.23b, the normalized interlayer shear modulus shows a decline for boron content <2.0%, and then remains almost constant at higher doping level, agreeing with the experimental results. The only distinction is that simulation results present a relatively larger decrease (~50%) in interlayer shear modulus than that of experiments (~20%), which is probably resulted from the discrepancy of system dimensions as well as the evaluation methods. For one thing, tiny cell sizes in simulations might enable the influence of varied interlayer spacing to be magnified on the overall mechanical behavior. For another, additional factors should be taken into account in the shear loading, such as the surface roughness and loading rates, which are eschewed in Raman experiments.

5.5.3 *Interlayer Slippage in Boron Doped Bilayer Graphene*

Given the poor van der Waals binding in multilayer graphene, the mechanical performance of multilayer graphene sheets would be greatly affected by the interlayer slippage [31, 32, 146, 147]. We thus perform nanoindentation tests with an AFM tip at the centre of the suspended bilayer graphene membrane, not only to characterize the interlayer slippage induced by the local shear deformation during indentation process, but also to clarify the effect of the slippage on the energy dissipation during loading/unloading cycle as depicted in Fig. 5.24a.

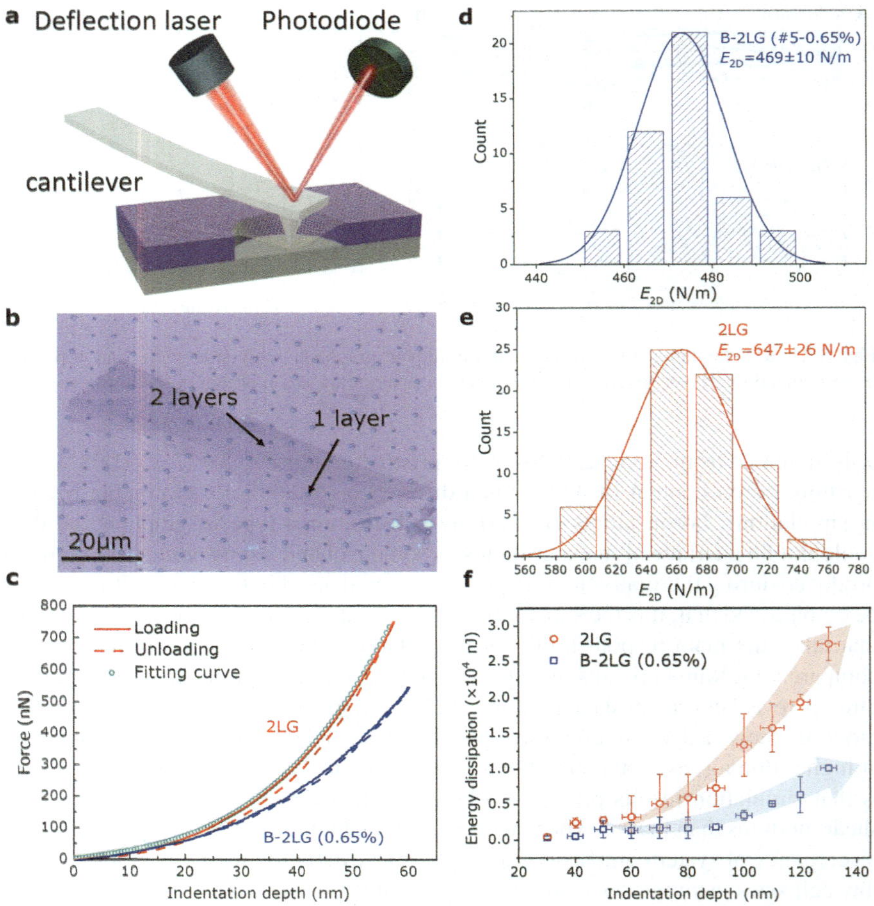

Fig. 5.24 **a** The schematic of the experimental setup for nanoindentation. **b** Optical image of monolayer and bilayer graphene sheets covering the patterned holes on the silicon substrate. **c** Experimental loading/unloading curves for 2LG and B-2LG and corresponding histograms of Young's modulus for **d** B-2LG and **e** 2LG, respectively. **f** Energy dissipations as a function of indentation depth for 2LG and B-2LG, respectively

The representative force-indentation curves for 2LG and boron-doped bilayer graphene (B-2LG) with boron content of ~2.4% are presented in Fig. 5.24c, and show a cubic dependence fitting the expression [48]:

$$F(\delta) = \pi \sigma_0 \delta + \frac{E_{2D}}{a^2} \delta^3 \tag{5.18}$$

where F is the loading force, δ is the indentation at the central point, σ_0 is the pre-tension accumulated in the sheet during the preparation procedure, E_{2D} is the two-dimensional Young's modulus and a is the radius. The results obtained from more than 30 drumheads yield histograms of E_{2D} as presented in Fig. 5.24d, e, in which $E_{2D} = 469 \pm 10$ N/m for B-2LG is lower than that of 2LG ($E_{2D} = 647 \pm 46$ N/m). The decreased trend could be attributed to the imperfect structure of graphene caused by the introduction of boron atoms [24, 148]. Furthermore, an apparent hysteresis loop related to the interlayer energy dissipation during a loading/unloading cycle was observed in Fig. 5.24c. Actually, indentation forces induced by AFM tip tend to cause the local shear deformation and lead to interlayer slippage. With increasing indentation depth, the interlayer slippage tends to dissipate more energy as shown in Fig. 5.24f. Compared to pristine 2LG, B-2LG exhibits a much lower energy dissipation as the unloading curve almost coincides with the loading curve. The decreased trend is ascribed to the relatively weaker interlayer coupling within a finite slippage distance. Moreover, the reduced energy dissipation by tuning the interlayer coupling seems to endow boron-doped graphene huge potential in multilayer graphene-based resonators as interlayer shear induced energy dissipation accounts for the critical loss mechanism on the Q-factors. The less energy dissipation per vibrational cycle enabled the graphene NEMS to extend its operational lifetime by performing near optimal capacity for a longer period of time [149].

5.6 Chapter Summary

Through the pressurized bulging test on bilayer graphene, we report the first measurement of the interlayer shear stress, which is as low as 0.04 MPa. Such a weak interlayer coupling easily induces the shear sliding between 2D layers and accounts for the breakdown of the classical plate theory. Despite good agreement of the measured Young's moduli with those reported in literature ($E_{graphene} > E_{hBN} > E_{MoS2}$), the bending rigidity is found to follow an opposite trend: $D_{graphene} < D_{hBN} < D_{MoS2}$ for multilayers with comparable thickness. We propose a modified formula for the bending rigidity that aids for the quantification of the interlayer slippage as well as the coupling between stretching and bending. The results imply that MoS_2 exhibits the highest interlayer coupling strength, while graphene layers hold the weakest interactions. Furthermore, through a joint study of the Raman shear modes, MD simulations and AFM nanoindentations, we unveiled that the introduction of boron atoms could effectively modulate the interlayer coupling of multilayer graphene. The interlayer shear modulus is demonstrated to be reduced by ~20%, which coincides fairly well with the enlarged interlayer spacing. The attenuated interlayer coupling in boron-doped multilayer graphene could greatly weaken the shear interactions during the slippage and hence aid the interpretation for the suppressed energy dissipation. Using the graphene scaffold, new 2D materials and 2D graphenic superlattices with fascinating properties can be created based on the delicate assembly and well-controlled chemical modification, which provides an attractive route to go beyond graphene.

References

1. Geim AK, Novoselov KS (2007) The rise of graphene. Nat Mater 6(3):183–191
2. Geim AK (2009) Graphene: status and prospects. Science 324(5934):1530–1534
3. Bonaccorso F, Colombo L, Yu G, Stoller M, Tozzini V, Ferrari AC et al (2015) Graphene, related two-dimensional crystals, and hybrid systems for energy conversion and storage. Science 347(6217):1246501
4. Li X, Cai W, An J, Kim S, Nah J, Yang D et al (2009) Large-area synthesis of high-quality and uniform graphene films on copper foils. Science 324(5932):1312–1314
5. Novoselov KS, Geim AK, Morozov SV, Jiang D, Zhang Y, Dubonos SV et al (2004) Electric field effect in atomically thin carbon films. Science 306(5696):666–669
6. Cao C, Mukherjee S, Howe JY, Perovic DD, Sun Y, Singh CV, Filleter T (2018) Nonlinear fracture toughness measurement and crack propagation resistance of functionalized graphene multilayers. Sci Adv 4:eaao7202
7. Bunch JS, Van der Zande AM, Verbridge SS, Frank IW, Tanenbaum DM, Parpia JM et al (2007) Electromechanical resonators from graphene sheets. Science 315(5811):490–493
8. Shahil KMF, Balandin AA (2012) Graphene-multilayer graphene nanocomposites as highly efficient thermal interface materials. Nano Lett 12(2):861–867
9. Nair R, Wu H, Jayaram P, Grigorieva I, Geim A (2012) Unimpeded permeation of water through helium-leak-tight graphene-based membranes. Science 335(6067):442–444
10. Dai Z, Liu L, Zhang Z (2019) Strain engineering of 2D materials: issues and opportunities at the interface. Adv Mater 31(45):1805417
11. Geim AK, Grigorieva IV (2013) Van der Waals heterostructures. Nature 499(7459):419–425
12. Novoselov KS, Mishchenko A, Carvalho A, Castro Neto AH (2016) 2D materials and van der Waals heterostructures. Science 353(6298):aac9439.
13. Liu Y, Weiss NO, Duan X, Cheng H-C, Huang Y, Duan X (2016) Van der Waals heterostructures and devices. Nat Rev Mater 1(9):16042
14. Lin Z, Huang Y, Duan X (2019) Van der Waals thin-film electronics. Nat Electron 2(9):378–388
15. Rogers GW, Liu JZ (2011) Graphene actuators: quantum-mechanical and electrostatic double-layer effects. J Am Chem Soc 133(28):10858–10863
16. Zhou M, Zhai Y, Dong S (2009) Electrochemical sensing and biosensing platform based on chemically reduced graphene oxide. Anal Chem 81(14):5603–5613
17. Liu C, Yu Z, Neff D, Zhamu A, Jang BZ (2010) Graphene-based supercapacitor with an ultrahigh energy density. Nano Lett 10(12):4863–4868
18. Hutchinson JW, Suo Z (1991) Mixed mode cracking in layered materials. Adv Appl Mech 29:63–191
19. Gao Y, Kim S, Zhou S, Chiu H-C, Nélias D, Berger C et al (2015) Elastic coupling between layers in two-dimensional materials. Nat Mater 14(7):714–720
20. Tan PH, Han WP, Zhao WJ, Wu ZH, Chang K, Wang H et al (2012) The shear mode of multilayer graphene. Nat Mater 11(4):294–300
21. Chiritescu C, Cahill DG, Nguyen N, Johnson D, Bodapati A, Keblinski P et al (2007) Ultralow thermal conductivity in disordered, layered WSe_2 crystals. Science 315(5810):351–353
22. Bertolazzi S, Brivio J, Kis A (2011) Stretching and breaking of ultrathin MoS_2. ACS Nano 5(12):9703–9709
23. Conley H, Lavrik NV, Prasai D, Bolotin KI (2011) Graphene bimetallic-like cantilevers: probing graphene/substrate interactions. Nano Lett 11(11):4748–4752
24. Koenig SP, Boddeti NG, Dunn ML, Bunch JS (2011) Ultrastrong adhesion of graphene membranes. Nat Nanotechnol 6(9):543–546
25. Zhang R, Ning Z, Xu Z, Zhang Y, Xie H, Ding F et al (2016) Interwall friction and sliding behavior of centimeters long double-walled carbon nanotubes. Nano Lett 16(2):1367–1374
26. Zhang RF, Ning ZY, Zhang YY, Zheng QS, Chen Q, Xie HH et al (2013) Superlubricity in centimetres-long double-walled carbon nanotubes under ambient conditions. Nat Nanotechnol 8(12):912–916

27. Song Y, Mandelli D, Hod O, Urbakh M, Ma M, Zheng Q (2018) Robust microscale superlubricity in graphite/hexagonal boron nitride layered heterojunctions. Nat Mater 17(10):894–899
28. Deng Z, Smolyanitsky A, Li Q, Feng X-Q, Cannara RJ (2012) Adhesion-dependent negative friction coefficient on chemically modified graphite at the nanoscale. Nat Mater 11(12):1032–1037
29. Lee C, Li Q, Kalb W, Liu X-Z, Berger H, Carpick RW et al (2010) Frictional characteristics of atomically thin sheets. Science 328(5974):76–80
30. Filleter T, McChesney JL, Bostwick A, Rotenberg E, Emtsev K, Seyller T et al (2009) Friction and dissipation in epitaxial graphene films. Phys Rev Lett 102(8):086102
31. Liu K, Yan Q, Chen M, Fan W, Sun Y, Suh J et al (2014) Elastic properties of chemical-vapor-deposited monolayer MoS_2, WS2, and their bilayer heterostructures. Nano Lett 14(9):5097–5103
32. Wei X, Meng Z, Ruiz L, Xia W, Lee C, Kysar JW et al (2016) Recoverable slippage mechanism in multilayer graphene leads to repeatable energy dissipation. ACS Nano 10(2):1820–1828
33. Ruoff R (2008) Calling all chemists. Nat Nanotechnol 3(1):10–11
34. Sun Z, James DK, Tour JM (2011) Graphene chemistry: synthesis and manipulation. J Phys Chem Lett 2(19):2425–2432
35. Liao L, Peng H, Liu Z (2014) Chemistry makes graphene beyond graphene. J Am Chem Soc 136(35):12194–12200
36. Wang G, Dai Z, Wang Y, Tan P, Liu L, Xu Z et al (2017) Measuring interlayer shear stress in bilayer graphene. Phys Rev Lett 119(3):036101
37. Wang G, Dai Z, Xiao J, Feng S, Weng C, Liu L et al (2019) Bending of multilayer van der Waals materials. Phys Rev Lett 123(11):116101
38. Wang G, Li X, Wang Y, Zheng Z, Dai Z, Qi X et al (2017) Interlayer coupling behaviors of boron doped multilayer graphene. J Phys Chem C 121(46):26034–26043
39. Ni Z, Wang H, Kasim J, Fan H, Yu T, Wu Y et al (2007) Graphene thickness determination using reflection and contrast spectroscopy. Nano Lett 7(9):2758–2763
40. Li H, Wu JMT, Huang X, Lu G, Yang J, Lu X et al (2013) Rapid and reliable thickness identification of two-dimensional nanosheets using optical microscopy. ACS Nano 7(11):10344–10353
41. Hao Y, Wang Y, Wang L, Ni Z, Wang Z, Wang R et al (2010) Probing layer number and stacking order of few-layer graphene by Raman spectroscopy. Small 6(2):195–200
42. Li XL, Qiao XF, Han WP, Lu Y, Tan QH, Liu XL et al (2015) Layer number identification of intrinsic and defective multilayered graphenes up to 100 layers by the Raman mode intensity from substrates. Nanoscale 7:8135–8141
43. Cai Q, Li LH, Yu Y, Liu Y, Huang S, Chen Y et al (2015) Boron nitride nanosheets as improved and reusable substrates for gold nanoparticles enabled surface enhanced Raman spectroscopy. Phys Chem Chem Phys 17(12):7761–7766
44. Li LH, Cervenka J, Watanabe K, Taniguchi T, Chen Y (2014) Strong oxidation resistance of atomically thin boron nitride nanosheets. ACS Nano 8(2):1457–1462
45. Lee C, Yan H, Brus LE, Heinz TF, Hone J, Ryu S (2010) Anomalous lattice vibrations of single- and few-layer MoS_2. ACS Nano 4(5):2695–2700
46. Li SL, Miyazaki H, Song H, Kuramochi H, Nakaharai S, Tsukagoshi K (2012) Quantitative raman spectrum and reliable thickness identification for atomic layers on insulating substrates. ACS Nano 6(8):7381–7388
47. Kim YA, Fujisawa K, Muramatsu H, Hayashi T, Endo M, Fujimori T et al (2012) Raman spectroscopy of boron-doped single-layer graphene. ACS Nano 6(7):6293–6300
48. Lee C, Wei X, Kysar JW, Hone J (2008) Measurement of the elastic properties and intrinsic strength of monolayer graphene. Science 321(5887):385–388
49. Lopez-Polin G, Gomez-Navarro C, Parente V, Guinea F, Katsnelson MI, Perez-Murano F et al (2015) Increasing the elastic modulus of graphene by controlled defect creation. Nat Phys 11(1):26–31

50. Gong L, Young RJ, Kinloch IA, Haigh SJ, Warner JH, Hinks JA et al (2013) Reversible loss of bernal stacking during the deformation of few-layer graphene in nanocompo sites. ACS Nano 7(8):7287–7294
51. Wang ZJ, Dong J, Cui Y, Eres G, Timpe O, Fu Q et al (2016) Stacking sequence and interlayer coupling in few-layer graphene revealed by in situ imaging. Nat Commun 7:13256
52. Huang M, Yan H, Chen C, Song D, Heinz TF, Hone J (2009) Phonon softening and crystallographic orientation of strained graphene studied by Raman spectroscopy. Proc Natl Acad Sci 106(18):7304–7308
53. He Y, Yang Y, Zhang Z, Gong Y, Zhou W, Hu Z et al (2016) Strain-induced electronic structure changes in stacked van der Waals heterostructures. Nano Lett 16(5):3314–3320
54. Neumann C, Reichardt S, Venezuela P, Drogeler M, Banszerus L, Schmitz M et al (2015) Raman spectroscopy as probe of nanometre-scale strain variations in graphene. Nat Commun 6:8429
55. Metzger C, Rémi S, Liu M, Kusminskiy SV, Castro Neto AH, Swan AK et al (2009) Biaxial strain in graphene adhered to shallow depressions. Nano Lett 10(1):6–10
56. Kis A, Jensen K, Aloni S, Mickelson W, Zettl A (2006) Interlayer forces and ultralow sliding friction in multiwalled carbon nanotubes. Phys Rev Lett 97(2):025501
57. Liu Z, Yang J, Grey F, Liu JZ, Liu Y, Wang Y et al (2012) Observation of microscale superlubricity in graphite. Phys Rev Lett 108(20):205503
58. Kim S, Yu J, van der Zande AM (2018) Nano-electromechanical drumhead resonators from two-dimensional material bimorphs. Nano Lett 18(11):6686–6695
59. Benameur MM, Gargiulo F, Manzeli S, Autes G, Tosun M, Yazyev OV et al (2015) Electromechanical oscillations in bilayer graphene. Nat Commun 6:8582
60. Ye F, Lee J, Feng PX (2017) Atomic layer MoS_2-graphene van der Waals heterostructure nanomechanical resonators. Nanoscale 9(46):18208–18215
61. Castellanos-Gomez A, Roldan R, Cappelluti E, Buscema M, Guinea F, van der Zant HS et al (2013) Local strain engineering in atomically thin MoS_2. Nano Lett 13(11):5361–5366
62. Yang S, Wang C, Sahin H, Chen H, Li Y, Li SS et al (2015) Tuning the optical, magnetic, and electrical properties of ReSe2 by nanoscale strain engineering. Nano Lett 15(3):1660–1666
63. Bao W, Miao F, Chen Z, Zhang H, Jang W, Dames C et al (2009) Controlled ripple texturing of suspended graphene and ultrathin graphite membranes. Nat Nanotechnol 4(9):562–566
64. Quereda J, San-Jose P, Parente V, Vaquero-Garzon L, Molina-Mendoza AJ, Agrait N et al (2016) Strong modulation of optical properties in black phosphorus through strain-engineered rippling. Nano Lett 16(5):2931–2937
65. Zang J, Ryu S, Pugno N, Wang Q, Tu Q, Buehler MJ et al (2013) Multifunctionality and control of the crumpling and unfolding of large-area graphene. Nat Mater 12(4):321–325
66. Chen PY, Sodhi J, Qiu Y, Valentin TM, Steinberg RS, Wang Z et al (2016) Multiscale graphene topographies programmed by sequential mechanical deformation. Adv Mater 28(18):3564–3571
67. Tapaszto L, Dumitrica T, Kim SJ, Nemes-Incze P, Hwang C, Biro LP (2012) Breakdown of continuum mechanics for nanometre-wavelength rippling of graphene. Nat Phys 8(10):739–742
68. Xie S, Tu L, Han Y, Huang L, Kang K, Lao KU et al (2018) Coherent, atomically thin transition-metal dichalcogenide superlattices with engineered strain. Science 359(6380):1131–1136
69. Xu Z, Buehler MJ (2010) Geometry controls conformation of graphene sheets: membranes, ribbons, and scrolls. ACS Nano 4(7):3869–3876
70. Cui X, Kong Z, Gao E, Huang D, Hao Y, Shen H et al (2018) Rolling up transition metal dichalcogenide nanoscrolls via one drop of ethanol. Nat Commun 9(1):1301–1307
71. Zhao J, Deng QM, Ly TH, Han GH, Sandeep G, Rummeli MH (2015) Two-dimensional membrane as elastic shell with proof on the folds revealed by three-dimensional atomic mapping. Nat Commun 6:8935–8940
72. Annett J, Cross GL (2016) Self-assembly of graphene ribbons by spontaneous self-tearing and peeling from a substrate. Nature 535(7611):271–275

73. Levy N, Burke S, Meaker K, Panlasigui M, Zettl A, Guinea F et al (2010) Strain-induced pseudo–magnetic fields greater than 300 tesla in graphene nanobubbles. Science 329(5991):544–547
74. Sanchez DA, Dai Z, Wang P, Cantu-Chavez A, Brennan CJ, Huang R et al (2018) Mechanics of spontaneously formed nanoblisters trapped by transferred 2D crystals. Proc Natl Acad Sci 115(31):7884–7889
75. Khestanova E, Guinea F, Fumagalli L, Geim AK, Grigorieva IV (2016) Universal shape and pressure inside bubbles appearing in van der Waals heterostructures. Nat Commun 7:12587
76. Reserbat-Plantey A, Kalita D, Han Z, Ferlazzo L, Autier-Laurent S, Komatsu K et al (2014) Strain superlattices and macroscale suspension of graphene induced by corrugated substrates. Nano Lett 14(9):5044–5051
77. Jiang Y, Mao J, Duan J, Lai X, Watanabe K, Taniguchi T et al (2017) Visualizing strain-induced pseudomagnetic fields in graphene through an hBN magnifying glass. Nano Lett 17(5):2839–2843
78. Dai Z, Hou Y, Sanchez DA, Wang G, Brennan CJ, Zhang Z et al (2018) Interface-governed deformation of nanobubbles and nanotents formed by two-dimensional materials. Phys Rev Lett 121(26):266101
79. Blees MK, Barnard AW, Rose PA, Roberts SP, McGill KL, Huang PY et al (2015) Graphene kirigami. Nature 524(7564):204
80. Sadd MH (2009) Elasticity: theory, applications, and numerics. Academic Press
81. Zhang DB, Akatyeva E, Dumitrica T (2011) Bending ultrathin graphene at the margins of continuum mechanics. Phys Rev Lett 106(25):255503
82. Wei Y, Wang B, Wu J, Yang R, Dunn ML (2013) Bending rigidity and Gaussian bending stiffness of single-layered graphene. Nano Lett 13(1):26–30
83. Lu Q, Arroyo M, Huang R (2009) Elastic bending modulus of monolayer graphene. J Phys D Appl Phys 42(10):102002
84. Zelisko M, Ahmadpoor F, Gao H, Sharma P (2017) Determining the Gaussian modulus and edge properties of 2D materials: from graphene to lipid bilayers. Phys Rev Lett 119(6):068002
85. Li H, Wang J, Gao S, Chen Q, Peng L, Liu K et al (2017) Superlubricity between MoS_2 monolayers. Adv Mater 29(27):1701474
86. Ribeiro-Palau R, Zhang CJ, Watanabe K, Taniguchi T, Hone J, Dean CR (2018) Twistable electronics with dynamically rotatable heterostructures. Science 361(6403):690–693
87. Hod O, Meyer E, Zheng Q, Urbakh M (2018) Structural superlubricity and ultralow friction across the length scales. Nature 563(7732):485–492
88. Wang D, Chen G, Li C, Cheng M, Yang W, Wu S et al (2016) Thermally induced graphene rotation on hexagonal boron nitride. Phys Rev Lett 116(12):126101
89. Woods C, Withers F, Zhu M, Cao Y, Yu G, Kozikov A et al (2016) Macroscopic self-reorientation of interacting two-dimensional crystals. Nat Commun 7:10800
90. Kelly BT (1981) Physics of graphite. Applied Science Publishers, London
91. Tapasztó L, Dumitrică T, Kim SJ, Nemes-Incze P, Hwang C, Biró LP (2012) Breakdown of continuum mechanics for nanometre-wavelength rippling of graphene. Nat Phys 8(10):739–742
92. Wang P, Gao W, Cao ZY, Liechti KM, Huang R (2013) Numerical analysis of circular graphene bubbles. J Appl Mech 80(4):040905
93. Mansfield EH (2005) The bending and stretching of plates. Cambridge University Press
94. Falin A, Cai Q, Santos EJG, Scullion D, Qian D, Zhang R et al (2017) Mechanical properties of atomically thin boron nitride and the role of interlayer interactions. Nat Commun 8:15815–15813
95. Kim SM, Hsu A, Park MH, Chae SH, Yun SJ, Lee JS et al (2015) Synthesis of large-area multilayer hexagonal boron nitride for high material performance. Nat Commun 6:8662–8672
96. Pan F, Wang G, Liu L, Chen Y, Zhang Z, Shi X (2019) Bending induced interlayer shearing, rippling and kink buckling of multilayered graphene sheets. J Mech Phys Solids 122:340–363
97. Han E, Yu J, Annevelink E, Son J, Kang DA, Watanabe K et al (2020) Ultrasoft slip-mediated bending in few-layer graphene. Nat Mater 19(3):305–309

98. Gao W, Huang R (2011) Effect of surface roughness on adhesion of graphene membranes. J Phys D Appl Phys 44(45):452001
99. Wei Y, Wang B, Wu J, Yang R, Dunn ML (2012) Bending rigidity and Gaussian bending stiffness of single-layered graphene. Nano Lett 13(1):26–30
100. Kudin KN, Scuseria GE, Yakobson BI (2001) C2F, BN, and C nanoshell elasticity from ab initio computations. Phys Rev B 64(23):235406
101. Shirasaki T, Derre A, Menetrier M, Tressaud A, Flandrois S (2000) Synthesis and characterization of boron-substituted carbons. Carbon 38(10):1461–1467
102. Cattelan M, Agnoli S, Favaro M, Garoli D, Romanato F, Meneghetti M et al (2013) Microscopic view on a chemical vapor deposition route to boron-doped graphene nanostructures. Chem Mater 25(9):1490–1495
103. Sheng Z-H, Gao H-L, Bao W-J, Wang F-B, Xia X-H (2012) Synthesis of boron doped graphene for oxygen reduction reaction in fuel cells. J Mater Chem 22(2):390–395
104. Tang YB, Yin LC, Yang Y, Bo XH, Cao YL, Wang HE et al (2012) Tunable band gaps and p-type transport properties of boron-doped graphenes by controllable ion doping using reactive microwave plasma. ACS Nano 6(3):1970–1978
105. Wang L, Sofer Z, Šimek P, Tomandl I, Pumera M (2013) Boron-doped graphene: scalable and tunable p-type carrier concentration doping. J Phys Chem C 117(44):23251–23257
106. Yen W-C, Medina H, Huang J-S, Lai C-C, Shih Y-C, Lin S-M et al (2014) Direct synthesis of graphene with tunable work function on insulators via in situ boron doping by nickel-assisted growth. J Phys Chem C 118(43):25089–25096
107. Peng Z, Ye R, Mann JA, Zakhidov D, Li Y, Smalley PR et al (2015) Flexible boron-doped laser-induced graphene microsupercapacitors. ACS Nano 9(6):5868–5875
108. Usachov DY, Fedorov AV, Petukhov AE, Vilkov OY, Rybkin AG, Otrokov MM et al (2015) Epitaxial B-graphene: large-scale growth and atomic structure. ACS Nano 9(7):7314–7322
109. Agnoli S, Favaro M (2016) Doping graphene with boron: a review of synthesis methods, physicochemical characterization, and emerging applications. J Mater Chem A 4(14):5002–5025
110. Jones LE, Thrower PA (1991) Influence of boron on carbon fiber microstructure, physical properties and oxidation behavior. Carbon 29(2):251–269
111. Wu X, Radovic LR (2005) Inhibition of catalytic oxidation of carbon/carbon composites by boron-doping. Carbon 43(8):1768–1777
112. Hishiyama Y, Inagaki M (2001) Lattice parameter changes in graphite with boron doping. Carbon 39:137–158
113. Panchakarla L, Govindaraj A, Rao C (2007) Nitrogen-and boron-doped double-walled carbon nanotubes. ACS Nano 1(5):494–500
114. Meyer JC, Geim AK, Katsnelson MI, Novoselov KS, Booth TJ, Roth S (2007) The structure of suspended graphene sheets. Nature 446(7131):60–63
115. Endo M, Hayashi T, Hong SH, Enoki T, Dresselhaus MS (2001) Scanning tunneling microscope study of boron-doped highly oriented pyrolytic graphite. J Appl Phys 90(11):5670–5674
116. Ta HQ, Perello DJ, Duong DL, Han GH, Gorantla S, Nguyen VL et al (2016) Stranski-Krastanov and Volmer-Weber CVD growth regimes to control the stacking order in bilayer graphene. Nano Lett 16(10):6403–6410
117. Lee S, Lee K, Zhong Z (2010) Wafer scale homogeneous bilayer graphene films by chemical vapor deposition. Nano Lett 10(11):4702–4707
118. Stephan O, Ajayan PM, Colliex C, Redlich P, Lambert JM, Bernier P et al (1994) Doping graphitic and carbon nanotube structures with boron and nitrogen. Science 266(5191):1683–1685
119. Dato A, Radmilovic V, Lee Z, Phillips J, Frenklach M (2008) Substrate-free gas-phase synthesis of graphene sheets. Nano Lett 8(7):2012–2016
120. Ci L, Song L, Jin CH, Jariwala D, Wu DX, Li YJ et al (2010) Atomic layers of hybridized boron nitride and graphene domains. Nat Mater 9(5):430–435
121. Zhan D, Sun L, Ni ZH, Liu L, Fan XF, Wang YY et al (2010) $FeCl_3$-based few-layer graphene intercalation compounds: single linear dispersion electronic band structure and strong charge transfer doping. Adv Funct Mater 20(20):3504–3509

122. Cancado LG, Jorio A, Ferreira EH, Stavale F, Achete CA, Capaz RB et al (2011) Quantifying defects in graphene via Raman spectroscopy at different excitation energies. Nano Lett 11(8):3190–3196
123. Zandiatashbar A, Lee GH, An SJ, Lee S, Mathew N, Terrones M et al (2014) Effect of defects on the intrinsic strength and stiffness of graphene. Nat Commun 5:3186
124. Ferrari AC, Basko DM (2013) Raman spectroscopy as a versatile tool for studying the properties of graphene. Nat Nanotechnol 8(4):235–246
125. Ferrari AC (2007) Raman spectroscopy of graphene and graphite: disorder, electron–phonon coupling, doping and nonadiabatic effects. Solid State Commun 143(1–2):47–57
126. Das A, Pisana S, Chakraborty B, Piscanec S, Saha SK, Waghmare UV et al (2008) Monitoring dopants by Raman scattering in an electrochemically top-gated graphene transistor. Nat Nanotechnol 3(4):210–215
127. Basko DM, Piscanec S, Ferrari AC (2009) Electron-electron interactions and doping dependence of the two-phonon Raman intensity in graphene. Phys Rev B 80(16):165413
128. Casiraghi C (2009) Doping dependence of the Raman peaks intensity of graphene close to the Dirac point. Phys Rev B 80(23):233407
129. Mohiuddin T, Lombardo A, Nair R, Bonetti A, Savini G, Jalil R et al (2009) Uniaxial strain in graphene by Raman spectroscopy: G peak splitting, Grüneisen parameters, and sample orientation. Phys Rev B 79(20):205433
130. Kawai S, Saito S, Osumi S, Yamaguchi S, Foster AS, Spijker P et al (2015) Atomically controlled substitutional boron-doping of graphene nanoribbons. Nat Commun 6:8098
131. Usachov DY, Fedorov AV, Vilkov OY, Petukhov AE, Rybkin AG, Ernst A et al (2016) Large-scale sublattice asymmetry in pure and boron-doped graphene. Nano Lett 16(7):4535–4543
132. Dai Z, Wang G, Zheng Z, Wang Y, Zhang S, Qi X et al (2019) Mechanical responses of boron-doped monolayer graphene. Carbon 147:594–601
133. Wang H, Feng M, Zhang X, Tan P-H, Wang Y (2015) In-phase family and self-similarity of interlayer vibrational frequencies in van der Waals layered materials. J Phys Chem C 119(12):6906–6911
134. Zhang X, Han WP, Wu JB, Milana S, Lu Y, Li QQ et al (2013) Raman spectroscopy of shear and layer breathing modes in multilayer MoS_2. Phys Rev B 87(11):115413
135. Zhao Y, Luo X, Li H, Zhang J, Araujo PT, Gan CK et al (2013) Interlayer breathing and shear modes in few-trilayer MoS_2 and WSe_2. Nano Lett 13(3):1007–1015
136. Cong C, Yu T (2014) Enhanced ultra-low-frequency interlayer shear modes in folded graphene layers. Nat Commun 5:4709
137. Nemanich RJ, Lucovsky G, Solin S (1977) Infrared active optical vibrations of graphite. Solid State Commun 23(2):117–120
138. Ling X, Liang L, Huang S, Puretzky AA, Geohegan DB, Sumpter BG et al (2015) Low-frequency interlayer breathing modes in few-layer black phosphorus. Nano Lett 15(6):4080–4088
139. Zhao W, Tan PH, Liu J, Ferrari AC (2011) Intercalation of few-layer graphite flakes with $FeCl_3$: Raman determination of Fermi level, layer by layer decoupling, and stability. J Am Chem Soc 133(15):5941–5946
140. Wu JB, Hu ZX, Zhang X, Han WP, Lu Y, Shi W et al (2015) Interface coupling in twisted multilayer graphene by resonant Raman spectroscopy of layer breathing modes. ACS Nano 9(7):7440–7449
141. Wu JB, Lin ML, Cong X, Liu HN, Tan PH (2018) Raman spectroscopy of graphene-based materials and its applications in related devices. Chem Soc Rev 47(5):1822–1873
142. Born M, Huang K, Lax M (1955) Dynamical theory of crystal lattices. Am J Phys 23(7):474–474
143. Wu JB, Zhang X, Ijas M, Han WP, Qiao XF, Li XL et al (2014) Resonant Raman spectroscopy of twisted multilayer graphene. Nat Commun 5:5309
144. Bosak A, Krisch M, Mohr M, Maultzsch J, Thomsen C (2007) Elasticity of single-crystalline graphite: inelastic X-ray scattering study. Phys Rev B 75(15):153408

145. Shen Y, Wu H (2012) Interlayer shear effect on multilayer graphene subjected to bending. Appl Phys Lett 100(10):101909
146. Lee C, Wei X, Li Q, Carpick R, Kysar JW, Hone J (2009) Elastic and frictional properties of graphene. Phys Status Solidi B Basic Solid State Phys 246(11):2562–2567
147. Xu L, Ma T-b, Hu Y-z, Wang H (2012) Molecular dynamics simulation of the interlayer sliding behavior in few-layer graphene. Carbon 50(3):1025–1032
148. Neek-Amal M, Peeters FM (2010) Nanoindentation of a circular sheet of bilayer graphene. Phys Rev B 81(23):235421
149. Kim SY, Park HS (2009) Multilayer friction and attachment effects on energy dissipation in graphene nanoresonators. Appl Phys Lett 94(10):101918

Chapter 6
Conclusions and Future Work

6.1 Conclusions

Interfacial mechanics of graphene not only attracts broad scientific interests but also represents a practical issue to developing a host of technological applications. In the preceding chapters, we have comprehensively studied different types of graphene-based interfaces, with particular emphasis on the experimental measurement of key parameters and characterization of nanoscale deformation mechanisms. The utmost original results achieved in this work are summarized as follows.

In Chap. 2, by stretching monolayer graphene on PMMA substrate, we uncover that the van der Waals interface is weak in the sense of shear and hence susceptible to slippage that limits the stress transfer efficiency from the substrate to the graphene sheet. The interfacial shear stress can be well modeled by a nonlinear shear-lag model, which governs the strain distribution in graphene as captured by Raman mapping. Significant degradation of interfacial stability has been observed under cyclic loading conditions, mainly due to the formation of buckles that impair the full contact at the interface. To this end, one effective solution is to modify the interfacial chemistry. It is demonstrated that the hydrogen bonding remarkably improves the interfacial shear strength up to 1.7 MPa, which approximately tripled that of pristine van der Waals (vdW) interface (0.6 MPa). Note that, the interfacial strengthening is compromised with structural damages caused by the chemical functionalization over the reinforcing efficiency of graphene, thus calling for an optimal design of interfaces to achieve the desired performance of graphene-based nanocomposites.

Going forward we have shown in Chap. 3 two different buckling modes of graphene embedded in PMMA matrix during thermal shrinkage. At relatively low strain range, the cooperative wrinkling mode is identified where graphene and PMMA deform coherently. Accordingly, the critical strain is determined by the interfacial stress transfer, depending on the lateral dimensions as well as the layer numbers of graphene sheets. The well-bonded interface endows the sandwich system with excellent mechanical recoverability under cyclic loading. A transition from wrinkling mode to buckle delamination is predicted at higher strain level. The debonding

G. Wang, *Characterization and Modification of Graphene-Based Interfacial Mechanical Behavior*, Springer Theses, https://doi.org/10.1007/978-981-15-8029-1_6

areas further act as the interfacial defects to accelerate the permanent failure at the interface which is detrimental to graphene-based flexible electronic devices.

Chapter 4 presents a combined Raman-AFM-bulging technique for measuring the interfacial properties between graphene and SiO_2 substrate. In contrast to the ultrastrong adhesion of graphene membranes, the frictional resistance is considerably weak and allows an easy slip at graphene/SiO_2 interface, in the manifestation of controllably growing shear zone around the bubble. Incorporating such a sliding boundary condition into the Hencky's model not only results in good agreement between theory and experiment, but also enables the estimation of the interfacial shear stress between graphene and SiO_2, averaged as 1.6 MPa. Further increasing the pressure induce the interfacial delamination of graphene, where the adhesion energy is obtained to be 0.19 J/m^2. However, the slip effect tends to cause the formation of radial buckles extending outwards from the bubble, which guide the delamination behavior and yield an irregular shape. In this case, the adhesion performance can be simply judged based on the aspect ratio of the bubble, behaving as a universal scaling.

With our understanding of the interfacial shear zone provided by Raman mapping and modified Hencky's model involving the sliding boundary, we turn our attention to the interlayer shear behavior between 2D material layers in Chap. 5. Through the Raman-AFM-bulging platform, we for the first time reveal an interlayer shear stress of 40 kPa for Bernal-stacked bilayer graphene. Such considerably low interlayer friction plays an important role in the mechanical response of multilayer 2D materials, even upon out-of-plane deformations. In particular, the anomalous scaling of bending rigidity is unveiled given the presence of interlayer shear, which violates the classic plate theory ($D \sim Et^3$). With the thickness-independent Young's modulus following the trend as expected ($E_{\text{graphene}} > E_{\text{hBN}} > E_{\text{MoS}_2}$), their bending rigidity exhibit a completely reversed trend ($D_{\text{graphene}} < D_{\text{hBN}} < D_{\text{MoS}_2}$) at comparable thickness. By comparing the experimental measurements to theoretical predictions, a quantitative description of interlayer coupling of 2D materials is proposed. Chemical strategy such as boron doping is demonstrated to attenuate interlayer coupling of multilayer graphene due to the enlarged interlayer spacing, which further provides an avenue for interface engineering on demand.

6.2 Novelty

1. **Experimentally**, we have developed a multiscale platform for mechanical characterization of a variety of 2D materials. Specifically, various loading conditions are realized such as uni-/biaxial stretching and compressing, as well as bending, which could mimic the complex stress states in practical applications. In situ Raman spectroscopy affords to monitor the local strains with a microscale resolution, even when 2D materials are incorporated in matrix. AFM technique

offers the nanoscale probing of not only surface morphology but also out-of-plane displacements. This experimental platform overcomes the technical bottleneck of manipulating and loading 2D materials, given their atomic thickness and planar nature. It also aids to tackle the challenges regarding complex mechanical responses that involve non-linearity, large deformation, instability and so forth. More importantly, the measured interfacial parameters (e.g. interfacial shear strength and adhesion) between different materials (e.g. 2D materials, flexible polymer, rigid semiconductor) will have practical implications for optimal design of 2D material-based composites and devices. One example is the first measurement of interlayer shear stress between graphene layers, which demonstrates the superlubric nature of graphene.

2. **Theoretically**, the key role of the shear deformations associated with the atomically smooth interfaces is incorporated into the continuum mechanics models. Previously, mathematic modeling typically assumes a clamped boundary for 2D material sheets when subjecting to the bulging or nanoindentation deformation; however, significant issues arise if we consider the subtle nature of the 2D material interfaces, which actually feature a weak resistance to tangential shear. The exploration into microscopic details of the deformation process involving the interfacial shear deformation hence provide a better understanding of nanomechanical performance of 2D materials. Furthermore, the modified mechanics model allows the accurate measurement of interfacial properties, which would also help to precisely characterize the intrinsic mechanical parameters of 2D materials since they interweave often.

6.3 Future Work

1. **Material system**: Although the mechanical exfoliation can produce high-quality 2D materials, it is still far from fulfilling the industrial requirement. From the perspective of applications, scalable manufacturing (e.g. CVD growth) of large-area 2D films is necessary [1]. The mechanical investigation of CVD graphene-based interfaces as well as its interplay with structural defects (e.g. grain boundaries) hence calls for further studies. In addition, 2D heterostructures assembled via vdW interactions have sparked immense interest in fields from physics to electronics [2–4]. A full understanding of these heterointerfaces is critical for the sophisticated construction and manipulation of vdW heterostructures at the fundamental atomic levels. Moreover, the twist angle as one of the crucial structural parameters in a vdW homo-/heterostructure, has emerged as a new degree of freedom for modulating its properties [5]. However, the twist angle dependence of interfacial properties still remains unexplored, which is yet crucially important for the device performance.
2. **Characterization technique**: In order to further improve the precision of mechanical characterization and reveal the structure–property relationship at the atomic level, state-of-art technology with higher spatial resolution is required,

such as scanning transmission electron microscope (STEM) [6, 7], scanning tunneling microscope (STM) [8, 9], tip-enhanced Raman spectroscopy (TERS) [10] and near-field infrared nanoscopy [11]. For example, sliding and twisting van der Waals layers with respect to each other gives rise to moiré structures [12], which can be clearly visualized by TEM or STM. In turn, the strain distribution and out-of-plane displacement in moiré patterns give direct information on the interlayer interactions in vdW heterostructures [5, 13]. In addition, despite its relative weakness, the vdW interlayer interaction between neighboring vdW layers can also induce significant structural reconstruction at the interface by favoring interlayer commensurability, which competes with the intralayer lattice distortion [6]. Consequently, a pattern of triangular Bernal-stacked domains separated by a network of stacking faults is generated. Mathematically, this domain boundary takes the form of a soliton that is finite in width but free to move, storing tensile or shear strains [7, 14]. In addition to abovementioned methods, the development of conductive atomic-force microscopy (cAFM) [15] and piezoresponse force microscopy (PFM) [16] techniques allows the direct observation of such strain solitons. Hence, we are able to gain more insights into their interaction with the interlayer coupling.

References

1. Lin L, Deng B, Sun J, Peng H, Liu Z (2018) Bridging the gap between reality and ideal in chemical vapor deposition growth of graphene. Chem Rev 118(18):9281–9343
2. Geim AK, Grigorieva IV (2013) Van der Waals heterostructures. Nature 499(7459):419–425
3. Novoselov KS, Mishchenko A, Carvalho A, Castro Neto AH (2016) 2D materials and van der Waals heterostructures. Science 353(6298):aac9439
4. Liu Y, Weiss NO, Duan X, Cheng H-C, Huang Y, Duan X (2016) Van der Waals heterostructures and devices. Nat Rev Mater 1(9):16042
5. Yao W, Wang E, Bao C, Zhang Y, Zhang K, Bao K et al (2018) Quasicrystalline 30 degrees twisted bilayer graphene as an incommensurate superlattice with strong interlayer coupling. Proc Natl Acad Sci 115(27):6928–6933
6. Yoo H, Engelke R, Carr S, Fang S, Zhang K, Cazeaux P et al (2019) Atomic and electronic reconstruction at the van der Waals interface in twisted bilayer graphene. Nat Mater 18(5):448–453
7. Alden JS, Tsen AW, Huang PY, Hovden R, Brown L, Park J et al (2013) Strain solitons and topological defects in bilayer graphene. Proc Natl Acad Sci 110(28):11256–11260
8. Kerelsky A, McGilly LJ, Kennes DM, Xian L, Yankowitz M, Chen S et al (2019) Maximized electron interactions at the magic angle in twisted bilayer graphene. Nature 572(7767):95–100
9. Zhang C, Chuu CP, Ren X, Li MY, Li LJ, Jin C et al (2017) Interlayer couplings, Moire patterns, and 2D electronic superlattices in MoS_2/WSe_2 hetero-bilayers. Sci Adv 3(1):1601459
10. Park KD, Raschke MB, Atkin JM, Lee YH, Jeong MS (2017) Probing bilayer grain boundaries in large-area graphene with tip-enhanced Raman spectroscopy. Adv Mater 29(7):1603601
11. Jiang LL, Shi ZW, Zeng B, Wang S, Kang JH, Joshi T et al (2016) Soliton-dependent plasmon reflection at bilayer graphene domain walls. Nat Mater 15(8):840–844
12. Woods CR, Britnell L, Eckmann A, Ma RS, Lu JC, Guo HM et al (2014) Commensurate–incommensurate transition in graphene on hexagonal boron nitride. Nat Phys 10(6):451–456

13. van Wijk MM, Schuring A, Katsnelson MI, Fasolino A (2014) Moire patterns as a probe of interplanar interactions for graphene on h-BN. Phys Rev Lett 113(13):135504
14. Butz B, Dolle C, Niekiel F, Weber K, Waldmann D, Weber HB et al (2014) Dislocations in bilayer graphene. Nature 505(7484):533–537
15. Weston A, Zou Y, Enaldiev V, Summerfield A, Clark N, Zólyomi V et al (2020) Atomic reconstruction in twisted bilayers of transition metal dichalcogenides. Nat Nanotechnol 1–6
16. McGilly LJ, Kerelsky A, Finney NR, Shapovalov K, Shih EM, Ghiotto A et al (2019) Seeing moire superlattices. arXiv preprint arXiv:1912.06629

The manufacturer's authorised representative in the EU is Springer Nature Customer Service Centre GmbH, Europaplatz 3, 69115 Heidelberg, Germany. If you have any concerns regarding our products, please contact ProductSafety@springernature.com

Printed and bound by CPI Group (UK) Ltd, Croydon, CR0 4YY
30/09/2026
02225515-0005